GREAT FORMULAS EXPLAINED

-

PHYSICS, MATHEMATICS, ECONOMICS

METIN BEKTAS

DEDICATION

This book is dedicated to my family.

CONTENTS

PART I: PHYSICS

PART II: MATHEMATICS

PART III: ECONOMICS

Part I: Physics

- **Intensity:**

Under ideal circumstances, sound or light waves emitted from a point source propagate in a spherical fashion from the source. As the distance to the source grows, the energy of the waves is spread over a larger area and thus the perceived intensity decreases. We'll take a look at the formula that allows us to compute the intensity at any distance from a source.

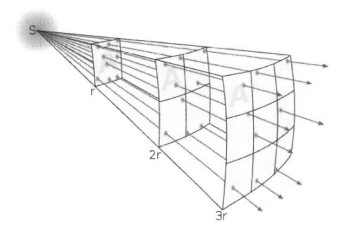

First of all, what do we mean by intensity? The intensity I tells us how much energy we receive from the source per second and per square meter. Accordingly, it is measured in the unit J per s and m^2 or simply W/m^2. To calculate it properly we need the power of the source P (in W) and the distance r (in m) to it.

$$I = P / (4 \cdot \pi \cdot r^2)$$

This is one of these formulas that can quickly get you hooked on physics. It's simple and extremely useful. In a later section you will meet the denominator again. It is the expression for the surface area of a sphere with radius r.

Before we go to the examples, let's take a look at a special intensity scale that is often used in acoustics. Instead of expressing the sound intensity in the common physical unit W/m^2, we convert it to its decibel value dB using this formula:

$$dB \approx 120 + 4.34 \cdot \ln(I)$$

with ln being the natural logarithm. For example, a sound intensity of $I = 0.00001$ W/m^2 (busy traffic) translates into 70 dB. This conversion is done to avoid dealing with very small or large numbers. Here are some typical values to keep in mind:

0 dB → Threshold of Hearing

20 dB → Whispering

60 dB → Normal Conversation

80 dB → Vacuum Cleaner

110 dB → Front Row at Rock Concert

130 dB → Threshold of Pain

160 dB → Bursting Eardrums

No onto the examples.

We just bought a P = 300 W speaker and want to try it out at maximal power. To get the full dose, we sit at a distance of only r = 1 m. Is that a bad idea? To find out, let's calculate the intensity at this distance and the matching decibel value.

$I = 300\ W / (4 \cdot \pi \cdot (1\ m)^2) \approx 23.9\ W/m^2$

$dB \approx 120 + 4.34 \cdot ln(23.9) \approx 134\ dB$

This is already past the threshold of pain, so yes, it is a bad idea. But on the bright side, there's no danger of the eardrums bursting. So it shouldn't be dangerous to your health as long as you're not exposed to this intensity for a longer period of time.

As a side note: the speaker is of course no point source, so all these values are just estimates founded on the idea that as long as you're not too close to a source, it can be regarded as a point source in good approximation. The more the source resembles a point source and the farther you're from it, the better the estimates computed using the formula will be.

Let's reverse the situation from the previous example. Again we assume a distance of r = 1 m from the speaker. At what power P would our eardrums burst? Have a guess before reading on.

As we can see from the table, this happens at 160 dB. To be able to use the intensity formula, we need to know the corresponding intensity in the common physical quantity W/m^2. We can find that out using this equation:

160 ≈ 120 + 4.34 · ln(I)

We'll subtract 120 from both sides and divide by 4.34:

40 ≈ 4.34 · ln(I)

9.22 ≈ ln(I)

The inverse of the natural logarithm ln is Euler's number e. In other words: e to the power of ln(I) is just I. So in order to get rid of the natural logarithm in this equation, we'll just use Euler's number as the basis on both sides:

$e^{9.22} \approx e^{ln(I)}$

10,100 ≈ I

Thus, 160 dB correspond to I = 10,100 W/m². At this intensity eardrums will burst. Now we can answer the question of which amount of power P will do that, given that we are only r = 1 m from the sound source. We insert the values into the intensity formula and solve for P:

10,100 = P / (4 · π · 1²)

10,100 = 0.08 · P

P ≈ 126,000 W

So don't worry about ever bursting your eardrums with a speaker or a set of speakers. Not even the powerful sound systems at rock concerts could accomplish this.

The intensity of the sunlight reaching earth is about I = 1400 W/m². Given that the distance between earth and sun is about r = 150,000,000,000 m, what is the sun's power output? To calculate this, we again have to solve the formula for P:

$$I = P / (4 \cdot \pi \cdot r^2)$$

$$P = I \cdot 4 \cdot \pi \cdot r^2$$

With this done, getting the result is simply a matter of plugging in the given data. This leads to:

$$P = 1400 \text{ W/m}^2 \cdot 4 \cdot \pi \cdot (150,000,000,000 \text{ m})^2$$

$$P \approx 4 \cdot 10^{26} \text{ W}$$

This value is almost beyond comprehension. In one second the sun gives off enough energy to satisfy the our current energy needs for the next 500,000 years. Unfortunately only very little of this energy actually reaches earth and only very little of that can be converted into useful energy.

In the next section we'll look at another radially propagating wave. But hopefully you noticed that physics and math does not need to be difficult. Some of the greatest calculations can be done with a handy formula and a few lines of writing.

- **Explosions:**

When a strong explosion takes place, a shock wave forms that propagates in a spherical manner away from the source of the explosion. The shock front separates the air mass that is heated and compressed due to the explosion from the undisturbed air. In the picture below you can see the shock sphere that resulted from the explosion of Trinity, the first atomic bomb ever detonated.

Using the concept of similarity solutions, the physicists Taylor and Sedov derived a simple formula that describes how the radius r (in m) of such a shock sphere grows with time t (in s). To apply it, we need to know two additional quantities: the energy of the explosion E (in J) and the density of the surrounding air D (in kg/m^3). Here's the formula:

$$r = 0.93 \cdot (E / D)^{0.2} \cdot t^{0.4}$$

Let's apply this formula for the Trinity blast.

In the explosion of the Trinity the amount of energy that was released was about 20 kilotons of TNT or:

$E = 84\ TJ = 84,000,000,000,000\ J$

Just to put that into perspective: in 2007 all of the households in Canada combined used about 1.4 TJ in energy. If you were able to convert the energy released in the Trinity explosion one-to-one into useable energy, you could power Canada for 60 years.

But back to the formula. The density of air at sea-level and lower heights is about $D = 1.25\ kg/m^3$. So the radius of the sphere approximately followed this law:

$r = 542 \cdot t^{0.4}$

After one second ($t = 1$), the shock front traveled 542 m. So the initial velocity was 542 m/s $\approx$ 1950 km/h $\approx$ 1210 mph. After ten seconds ($t = 10$), the shock front already covered a distance of about 1360 m $\approx$ 0.85 miles.

How long did it take the shock front to reach people two miles from the detonation? Two miles are approximately 3200 m. So we can set up this equation:

$3200 = 542 \cdot t^{0.4}$

We divide by 542:

$5.90 \approx t^{0.4}$

Then take both sides to the power of 2.5:

$t \approx 85\ s \approx 1$ *and 1/2 minutes*

Let's look at how the different parameters in the formula impact the radius of the shock sphere:

- If you increase the time sixfold, the radius of the sphere doubles. So if it reached 0.85 miles after ten seconds, it will have reached 1.7 miles after 60 seconds. Note that this means that the speed of the shock front continuously decreases.

For the other two parameters, it will be more informative to look at the initial speed v (in m/s) rather the radius of the sphere at a certain time. As you noticed in the example, we get the initial speed by setting t = 1, leading to this formula:

$$v = 0.93 \cdot (E / D)^{0.2}$$

- If you increase the energy of the detonation 35-fold, the initial speed of the shock front doubles. So for an atomic blast of 20 kt · 35 = 700 kt, the initial speed would be approximately 542 m /s · 2 = 1084 m/s.

- The density behaves in the exact opposite way. If you increase it 35-fold, the initial speed halves. So if the test were conducted at an altitude of about 20 miles (where the density is only one thirty-fifth of its value on the ground), the shock wave would propagate at 1084 m/s

Another field in which the Taylor-Sedov formula is commonly applied is astrophysics, where it is used to model Supernova explosions. Since the energy released in such explosions dwarfs all atomic blasts and the surrounding density in space is very low, the initial expansion rate is extremely high.

- **Mach Cone:**

When an object moves faster than the speed of sound, it will go past an observer before the sound waves emitted by object do. The waves are compressed so strongly that a shock front forms. So instead of the sound gradually building up to a maximum as it is usually the case, the observer will hear nothing until the shock front arrives with a sudden and explosion-like noise.

Geometrically, the shock front forms a cone around the object, which under certain circumstances can even be visible to the naked eye (see image below). The great formula that is featured in this section deals with the opening angle of said cone. This angle, symbolized by the Greek letter θ, is also indicated in the image.

All we need to compute the mach angle θ is the velocity of the object v (in m/s) and speed of sound c (in m/s):

sin θ = c / v

Let's turn to an example.

A jet fighter flies with a speed of v = 500 m/s toward its destination. It flies close to the ground, so the speed of sound is approximately c = 340 m/s. This leads to:

sin θ = 340 / 500 = 0.68

θ = arcsin(0.68) ≈ 43°

In the picture above the angle is approximately 62°. How fast was the jet going at the time when the picture was taken? We'll set the speed of sound to c = 340 m/s and insert all the given data into the formula:

sin 62° = 340 / v

0.88 = 340 / v

Obviously we need to solve for v. To do that, we first multiply both sides by v. This leads to:

0.88 · v = 340

Dividing both sides by 0.88 results in the answer:

v = 340 / 0.88 ≈ 385 m/s ≈ 1390 km/h ≈ 860 mph

There's a formula which can serve as a helpful supplement to calculating the mach angle (or as in the second example, the speed of the object). The air temperature varies with height. As we go higher, the temperature gets lower. On average there's a 6 °C drop in temperature for every kilometer additional altitude. So when it's 20 °C on the ground, we can expect about - 40 °C at a height of 10 km.

Why should we care about temperature here? After all, it's not an input for the formula. That's true, however, the speed of sound depends on temperature. The hotter it is, the faster sound waves propagate. The formula below can be used to approximate the speed of sound c (in m/s) from the air temperature (in °C):

$$c = 331 \cdot \text{sq root} (1 + T / 273)$$

At a temperature T = 20 °C, an average autumn day, sound moves with c ≈ 343 m/s. When we go 10 km up, which is roughly the cruising altitude of large planes, the temperature drops to T = - 40 °C and with it the speed of sound to c ≈ 306 m/s. Here the sound waves move circa 10 % slower!

- **Reverberation:**

When you clap your hands in a very large room, you can notice the sound persisting for a short time. The reason for that is that the sound waves are reflected back and forth between the walls, creating a large number and a complex pattern of echos.

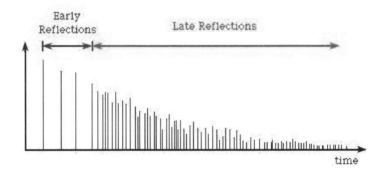

At the end of the 19th century, Wallace Clement Sabine empirically studied the reverberation time at Harvard University and derived a handy approximation formula for it. By reverberation time we mean the time it takes for the sound to decay by 60 dB.

It depends on four quantities: the volume of the room V (in m^3), the total surface area of the room A (in m^2), the absorption coefficient of the surfaces a (dimensionless) and finally the absorption coefficient of air b (dimensionless). From these we can get an estimate for the reverberation time T (in s) using Sabine's formula, or rather, a slightly modified version of it:

$T = 0.16 \cdot V / (A \cdot a + V \cdot b)$

For common brickwork and plaster walls the absorption coefficient is about a = 0.03, for wood a = 0.3 and for acoustic tiles it can go as high as a = 0.8. As for the air absorption coefficient, it is roughly b = 0.02 at 50 % humidity.

You are in an empty rectangular hall with the dimensions 30 m by 30 m by 5 m. We assume the absorption coefficients to be a = 0.03 and b = 0.02. What is the reverberation time for this hall?

First we need to calculate the the volume:

$$V = 30 \ m \cdot 30 \ m \cdot 5 \ m = 4500 \ m^3$$

and the total surface area:

$$A = 2 \cdot (30 \ m \cdot 30 \ m + 30 \ m \cdot 5 \ m + 30 \ m \cdot 5 \ m)$$

$$A = 2400 \ m^2$$

Now we can turn to Sabine's formula:

$$T = 0.16 \cdot 4500 \ m^3 / (2400 \ m^2 \cdot 0.03 + 4500 \ m^3 \cdot 0.02)$$

$$T \approx 4.4 \ s$$

Note that for the units to check out, the constant 0.16 must have the unit s/m. This has been left out of the general formula and the example for simplicity.

If the computed reverberation time seems too high to you, remember that it is reduced significantly once the hall is filled with equipment and people (which is usually the case).

How does the reverberation time vary with the room dimensions and absorption coefficients?

- Here we cannot simply look at the volume and surface area separately since they depend on each other. The surface area grows approximately proportional to the 2/3 power of the volume, which means that overall, the reverberation time will grow with the third root of the volume for small rooms and reach a limiting value for larger rooms. This limiting value is 0.16 / b or about 8 s at 50 % humidity. No matter how big the room, the reverberation time cannot go beyond that.

In the picture below you can see the variation of reverberation time with volume and how the curve flattens out as the volume increases and the limiting value is approached.

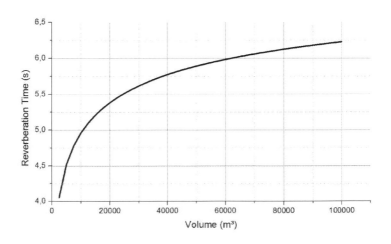

- If the absorption coefficients increase, the reverberation time decreases. If, for example, we were to cover the walls of the hall with high quality acoustic tiles, the reverberation time would drop to a mere $T \approx 0.4$ s.

Keep in mind that the formula just delivers a useful first estimate. In reality, the process of sound reflection depends in a very complex way on the specific geometry of a room. Two halls with the same volume, surface area and absorption coefficients can produce very different reverberation times.

- **Doppler:**

Have you ever listened carefully as a police car swooshed by? If yes, then you probably noticed this strange sound effect. When the car approached, the pitch of the siren was relatively high. At the moment it passed you, the pitch changed quickly to a lower tone, where it remained while the car drove off. What happened?

If a source of sound is moving, the sound waves emitted in the direction of travel are compressed. This means that the wavelength gets shorter and accordingly the frequency (pitch) higher. So as long as the police car is approaching you, you'll receive compressed sound waves from the siren.

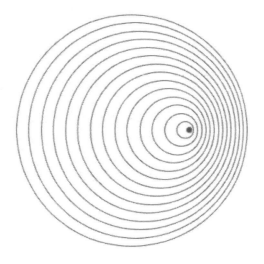

The opposite happens to sound waves that are emitted against the direction of travel. Their wavelength gets longer and the frequency lower. These are the sound waves you receive when the police car has passed you and is driving off. Only at the exact moment the car is passing you do you hear the original tone of the siren.

This effect is not limited to sound waves, it also occurs in the case of light. Stars and galaxies moving towards or away from earth are observed with a slight change in light frequency (color). These red- and blue-shifts are commonly used by astronomers to determine radial velocities.

Let's get to the formula. We need three quantities as inputs: the original frequency of the source f (in Hz), the velocity at which it approaches the observer v (in m/s) and the speed of the waves c (in m/s). The observed frequency is:

f' = f / (1 - v / c)

A car emitting sound at the chamber pitch f = 440 Hz approaches you with v = 36 m/s. The speed of sound is about c = 340 m/s. At what frequency do you perceive the sound?

f' = 440 Hz / (1 - 36 / 340) ≈ *492 Hz*

So the Doppler effect shifted the tone two half-notes from the original note A to note B. That is certainly noticeable.

Here's how varying the inputs will change the frequency perceived by the observer:

- If the emitted frequency doubles, so does the perceived frequency. So 440 Hz · 2 = 880 Hz would change into 492 Hz · 2 = 984 Hz. Again, this corresponds to a change in pitch of about two half-tones.

- If the speed of the source increases, the perceived frequency increases as well, that is, the tone will be at an even higher pitch. At 72 m/s the frequency would have risen to 558 Hz, which is about four half-tones above the original pitch (note C#).

Up to this point we only looked at approaching sound sources. In the opposite case, the sound source moving away from the observer, we need to replace the minus-sign in the formula with a plus-sign to arrive at the correct result.

Again the car is emitting sound at the chamber pitch f = 440 Hz, but this time it moves away from you with v = 36 m/s. What frequency will you perceive then?

f' = 440 Hz / (1 + 36 / 340) ≈ 398 Hz

The shift again was about two half-tones, but this time to a lower pitch (from note A to note G).

When the speed of the source becomes equal to or greater than the speed of the waves it emits, the first formula will not work anymore. Obviously in such a case, you won't hear any sound from the approaching source as it will reach your position before the waves do. We already had a look at this situation in the section "Mach Cone".

- **Hurricanes:**

In this section we are going to do just what the title says, that is compute hurricanes. The great formula that accomplishes this, called Rankine formula, is very little known among physicists and mathematicians, most are not aware of its existence. But that doesn't make it any less useful.

One of the most important quantities that is used to characterize a hurricane, aside from the size, is the pressure difference p (usually in millibars, in short: mb) between the center and the surrounding of the hurricane. Air always flows from high to low pressure and thus, when an area of low pressure forms, air starts flowing towards it. Because of earth's rotation, the resulting flow is not direct. The air rather circulates around and into this region of low pressure. The greater the pressure difference, the more violent the movement of air will be.

For starters, we will assume this pressure difference to be constant over the life of a hurricane. At a later point we will relax this condition, allowing the calculations to include strengthening and weakening hurricanes. But for now, we only care about two quantities: the distance from an observer to the center of the storm r (any unit of length will do as long as we are consistent) and the wind speed v at this distance.

The Rankine formula states that this expression is conserved as the hurricane changes position:

$$v \cdot r^{0.6} = \text{constant}$$

Our strategy will be: first we use current data (a distance and a wind speed) to compute the constant, then we are able to get an estimate for the wind speed at any distance. Note that this equation tells us that when we triple the distance to the center of the hurricane, the wind speed halves.

A hurricane is approaching and according to TV reports it is currently about 600 miles away from our town. Current wind speeds are about 20 mph. From the projected path we can deduce that the hurricane will come as close as 100 miles. What is the maximum wind speed v we can expect?

First we determine the constant using the current data:

$20 \cdot 600^{0.6} \approx 930$

Now we can set up an equation for the maximum wind speed. Since we inputted the speed in mph, the result will be in the same unit.

$v \cdot 100^{0.6} \approx 930$

$v \cdot 16 \approx 930$

$v \approx 58 \ mph$

Simple as that. But remember that we assumed the hurricane to be of constant strength during its approach. If this is not the case, we need to include the pressure difference in our calculations, which is what we will do now.

In case of hurricanes of changing strength, the pressure difference p appears as a variable in the Rankine formula. This makes things a little harder, but luckily not by much.

$v \cdot r^{0.6} / sq \ root \ (p) = constant$

Let's turn to an example. We stick to the strategy: first determine the constant using current data (a distance, a wind speed and a pressure difference), then we can calculate the wind speed at any distance and pressure difference.

Again the approaching hurricane is 600 miles away with current wind speeds of 20 mph. The pressure difference between the center and surroundings of the hurricane at this point is about 60 mb. During its approach, it will come as close as 100 miles and is expected to strengthen to 80 mb. What is the maximum wind speed v we can expect?

First we determine the constant:

$20 \cdot (600)^{0.6} / sq\ root\ (60) \approx 120$

Now let's find the maximum wind speed:

$v \cdot (100)^{0.6} / sq\ root\ (80) \approx 120$

$v \cdot 1.8 \approx 120$

$v \approx 67\ mph$

It is important to note that all of the equations only hold true outside the eye of the storm (which is usually about 20 to 40 miles in diameter). The maximum wind speed in a hurricane is reached at the wall of the eye. Inside the eye wind speeds drop sharply. It is so to speak "the calm within the storm" and can make for a quite eerie experience.

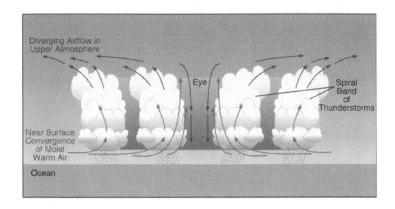

We'll draw one last conclusion before moving on. The size of the eye is more or less a constant. This implies that the maximum wind speed within a hurricane grows with the square root of the pressure difference. So if the pressure difference quadruples, the maximum wind speed will approximately double. Real-world data confirms this conclusion within acceptable boundaries. As an estimate for the maximum wind speed in a hurricane you can use this formula:

$v(max) \approx 16 \cdot sq\ root(p)$

The result is in mph. For a category four hurricane ($p = 80$ mb) we can thus expect maximum wind speeds of 140 mph.

- **Flow:**

Oil has become the blood of the world. We would need to give up many of the luxuries we unfortunately often take for granted if the supply of oil stopped. To ensure that the flow never stops, a vast network of pipelines has been built, transporting the precious fossil fuel day by day over distances of thousands of miles. In this section we will take a look at the flow of liquids and gases within pipes.

The formula we are focusing on here is called Hagen-Poiseuille law. It allows the calculation of the volume flow rate F (in m^3/s) in pipes. Which quantities do we need to accomplish that? We certainly need the dimensions of the pipe, or to be more specific, the radius r (in m) and the total length l (in m).

In the previous section we stated that in order for air (or any gas or liquid for that matter) to flow, we need to have a pressure difference p (here measured in Pascal $= Pa = N/m^2$). Thus, this quantity will also of importance in this case.

Last but not least, we need an additional quantity to characterize the fluid in the system. Given the same pipe dimensions and pressure difference, air will certainly flow at a very different rate than water or oil. So to compute the flow rate, we also need the so called dynamic viscosity μ (in Pa s).

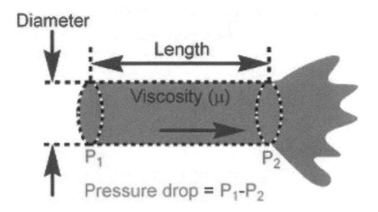

Now we're ready to state the formula:

$$F = \pi \cdot r^4 \cdot p / (8 \cdot \mu \cdot l)$$

A local pipeline with the dimensions r = 1 m and l = 25,000 m is used to transport oil (μ = 0.25 Pa s) to a city. To do that, a pump creates a pressure difference of p = 5000 Pa. What is the resulting volume flow rate?

We simply plug in the values into the formula. I will leave out the units for more clarity. Rest assured, they'll check out, Hagen and Poiseuille made sure of that.

$F = \pi \cdot 1^4 \cdot 5000 / (8 \cdot 0.25 \cdot 25,000)$

$F \approx 0.3 \ m^3/s \approx 1130 \ m^3/h$

The city is not satisfied with the flow rate. It asks us to increase this to 2500 m³/h. What pressure difference do we need to apply for that? First we have to make sure to use the correct units, which means the flow rate must be in m³/s.

2500 m³/h ≈ 0.69 m³/s

Now we can set up an equation using the formula and the values we know. This leads to:

0.69 = π · 1⁴ · p / (8 · 0.25 · 25,000)

0.69 ⁻ 0.000063 · p

p ≈ 11,000 Pa

We obviously need a much stronger pump to satisfy the city's demand. A word regarding conversion: you multiply m³/s by 3600 to get to m³/h and you divide m³/h by 3600 to get to m³/s. When looking at flow, this is a conversion you'll have to do quite often.

Let's take a look at how varying the input quantities affects the resulting volume flow rate:

- There is a very strong variation with pipe radius. If you double the radius, the volume flow increases sixteen-fold. The pipeline from the latter example would transport 2500 m³/h · 16 = 40,000 m³/h if the radius were 2 m.

- If you double the length of the pipeline (or the viscosity of the liquid), the volume flow rate halves. So going from 25 km to 50 km would reduce the flow rate from 2500 m³/h to 1250 m³/h if all other quantities remained the same.

- If you double the pressure difference, the volume flow rate doubles as well. So there's a simple proportional relationship between the two, which makes altering the flow by reducing or increasing pump power rather straight-forward.

I hope this bit helped you to appreciate the mathematics involved in fluid motion. It is certainly a very rich and interesting field that is worth going into. So next time you browse on Amazon, be sure to look for introductory books on this topic.

- **Traffic:**

We just talked about the flow of fluids, but I can think of at least one more thing that can (and sometimes just won't) flow and that is car traffic. Mathematicians have come up with dozens of models to simulate car traffic in order to make the whole system safer and more efficient. Some models have proven to be quite successful in this but in the end, there's always one important factor they can't get right: the unexpected and irrational behavior of people. Anger, frustration, stress - you can't put that into numbers.

Still it's worthwhile to look at some basics of car traffic mathematics. It has brought forth some neat formulas and useful conclusions. For the formulas, we will need three quantities: the flow rate F (in cars per hour), the velocity of the flow v (in miles per hour) and the traffic density D (in cars per mile).

There's a reason why I wrote "miles per hour" instead of the short form "mph". I did that because we can derive a very important formula by just looking at the units. Suppose we multiply the traffic density D with the flow velocity v. What is the unit of the resulting quantity? Well, if we multiply "cars per mile" with "miles per hour" we obviously get the unit "cars per hour" as a result, since the miles cancel each other out. This is just the unit of the traffic flow F. Thus:

$$F = D \cdot v$$

So if there are D = 40 cars per mile and the average speed is v = 50 mph, then the resulting traffic flow is F = 2000 cars per hour. Simple as that. There are two more formulas you should be aware of. As of now, we treated the traffic density and speed as independent quantities. But as you know from experience, this is not necessarily true. We know that as the traffic density grows, the traffic slows down. So there must be a relationship between the two.

Observation of traffic has shown that indeed there is and that in good approximation the average velocity decreases linearly with density. If we denote the free-flow velocity by u and the maximum density by M, then this formula provides a good estimate between traffic speed v and density D:

$$v = u \cdot (1 - D / M)$$

One word about the new quantities: the free-flow velocity is the speed that drivers choose when the road is almost empty. Accordingly, it is usually close to the speed limit. As for the maximum traffic density, it is usually around 300 cars per mile and lane, which corresponds to bumper to bumper traffic.

Let's do a quick example before looking at the interesting consequences of this relationship.

Observations have shown that on a certain one-lane road the free-flow velocity is u = 50 mph and the maximum density M = 300 cars per mile. We estimate that the current average velocity is about v = 20 mph. What is the current traffic density and flow rate?

First we'll use the second formula to determine the traffic density D from the given data:

20 = 50 · (1 - D / 300)

Divide by 50 and subtract 1 from both sides:

$0.4 = 1 - D / 300$

$- 0.6 = - D / 300$

Finally multiply by -300:

$D = 180 \text{ cars/mile}$

Now that we also know the density, we can easily compute the traffic flow rate F from the first formula:

$F = 20 \text{ mph} \cdot 180 \text{ cars/mile} = 3600 \text{ cars/hour}$

If we plug the relationship between the density and speed into the formula for traffic flow, we can see that the traffic flow varies in a parabolic fashion with the density. At low densities, when the road is almost empty, the traffic flow increases as the density grows. However, at a specific density the traffic flow reaches a maximum value and decreases after that. The exciting conclusion: for every road there's a maximum flow rate, which we call the capacity.

The derivation of the formula for the capacity C (in cars per hour) requires a bit of calculus, so we skip the derivation and go right to the formula. It includes only two quantities and those we know already: the free-flow velocity u (in mph) and the maximum traffic density M (in cars per mile).

$C = 0.25 \cdot u \cdot M$

The calculations also show that this maximum flow rate is always reached at half the maximum density.

----

*We go back to the one-lane road from the previous example
with u = 50 mph and M = 300 cars per mile. We computed
that at the current time the traffic flow rate is 3600 cars per
hour. How much higher can this go? What is the maximum
flow rate on this road?*

$C = 0.25 \cdot 50\ mph \cdot 300\ cars\ per\ mile$

$\approx 3750\ cars\ per\ hour$

*This flow rate will occur when the density drops from the
current D = 180 cars per mile to D = 300 / 2 = 150 cars per
mile. Any increase in density from the current value will
only lower the traffic flow rate*

*In the picture below you can see the theoretical relationship
between the density and flow rate for this one-lane road.*

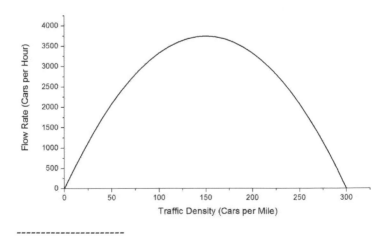

But remember that we are dealing with a large number of
people here experiencing a large number of different

emotions while driving. All we can hope for are good approximations. So take all of the formulas and results with a plus / minus ten percent or so accuracy. As long as people don't act like the rational beings traffic scientists and economists would like them to be, we have to settle for that.

- **Gravity:**

Every object that is in the proximity of other objects experiences a pull toward those. This is one of the most fundamental laws that exist in Physics. This pull keeps you connected to earth, earth connected to the sun, the sun connected to the Milky-Way and the Milky-Way connected to the local galaxy cluster. You know it by the name of gravity.

The formula describing this fundamental force was included by Newton in his book Principa, published in 1686. It relies upon three quantities: the mass of one object m (in kg), the mass of another object M (in kg) and the distance between them d (in m), measured from center to center. On top of that, the formula includes a constant, called the gravitational constant $G = 6.67 \cdot 10^{-11}$ N $(m/kg)^2$.

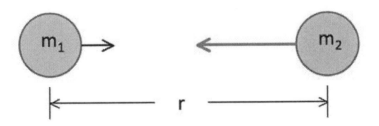

The formula for the gravitational force (in N) is:

$F = G \cdot m \cdot M / r^2$

Let's look at one example on how to apply this formula:

The mass of an average adult is about m = 75 kg. At any moment, the person experiences the gravitational pull of earth with its mass of M = 5.97 · 10²⁴ kg. The distance of a person to earth's core is about r = 6,370,000 m. Given this, what is the gravitational force acting on this person? We apply the formula:

$F = G \cdot 80 \ kg \cdot 5.97 \cdot 10^{24} \ kg \ / \ (6,370,000 \ m)^2$

$F \approx 785 \ N$

Let's look at how changing the inputs impacts the gravitational force, before turning to a neat simplification.

- If we double the mass of one of the objects, the gravitational force doubles as well. So a 160 kg person would experience a force of roughly 785 N · 2 = 1570 N. The same force would act on a 80 kg person on a planet twice the mass of earth.

- If we double the distance between the objects, the force decreases by a factor of four. So if our average adult were in a space station located at a distance 6370 km · 2 = 12740 km from earth's core, the gravitational force would drop to about 785 N : 4 = 195 N. As you can see, the drop is sharp, which is why astronauts orbiting earth are basically weightless.

When we calculate gravitational forces on earth's surface, we don't need to go through the hassle of dealing with very small (gravitational constant) or very large (earth's mass) numbers every time. As you can verify using the above formula, the force on a 1 kg mass located on earth's surface is:

$$g \approx 9.81 \text{ N}$$

As common in physics, I abbreviated this special value by g. Scientists call this the gravitational acceleration. With this number, we can write the law of gravitation as such:

$$\mathbf{F} = \mathbf{m} \cdot \mathbf{g}$$

with m being the mass of the object on earth's surface. Why is g called gravitational acceleration? You might remember Newton's second law, which states that the force of inertia is the product of mass and acceleration:

$$F = m \cdot a$$

When you drop a body, the movement is caused by the gravitational pull. So we insert the formula for gravitation for F to determine the resulting acceleration:

$$m \cdot g = m \cdot a$$

$$g = a$$

So g is not only the force a 1 kg object experiences on earth's surface, it is also the acceleration any object that is dropped is subject to. Certainly an important value to keep in mind.

You might say to yourself: that can't be right. When you drop a feather, it doesn't accelerate as fast as a stone. The thing is: it does ... in vacuum. Initially the feather experiences exactly the same acceleration as the stone, 9.81 m/s^2. It is only the presence of air that makes them fall differently. If you let both the feather and the stone drop in a vacuum tube, both reach the bottom at the same time. Make sure to check out this experiment, it's a real eye-opener to see a feather literally drop like a stone.

One quick note about gravitation: it is the determining factor for the universe on a large scale as well as on a human scale. But when things get smaller, that is, when we go to the realm of molecules, atoms or even sub-atomic particles, gravity becomes insignificant. On this level, the electromagnetic, strong and weak force take over. Atoms don't care for gravity.

- **Range:**

When you throw an object, gravity will force it on a parabolic path. One fundamental question that arises here is: given that you threw this object at a velocity of v (in m/s) and an angle of θ (read "theta", in °), how far will it go? To answer this, you need to describe the trajectory in mathematical form and intersect this curve with the ground. Doing that results in this very useful formula for the range R (in m):

$$R = v^2 \cdot \sin (2 \cdot \theta) / g$$

with g being the gravitational acceleration. Let's do a simple and straight-forward application.

A Dolphin jumps out of the water at a velocity of v = 5.5 m/s and an angle of θ = 70° as shown in the image. What distance will it cover before hitting the water? Remember that the gravitational acceleration is g = 9.81 m/s².

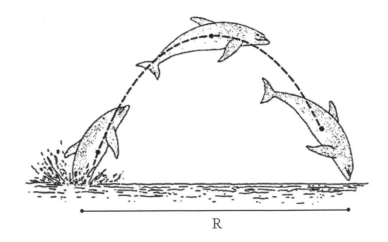

R

$R = (5.5 \text{ m/s})^2 \cdot sin(140°) / 9.81 \text{ m/s}^2 \approx 2 \text{ m}$

If the dolphin wants to get far, the best choice is to jump at an angle of $\theta = 45°$. This would increase the range to:

$R = (5.5 \text{ m/s})^2 \cdot sin(90°) / 9.81 \text{ m/s}^2 \approx 3.1 \text{ m}$

which is the maximum range at the given velocity.

Let's look at how changing the quantities that determine the trajectory impact the resulting range:

- If the velocity is doubled, the range increases fourfold. So by jumping at 11 m/s rather than 5.5 m/s, the dolphin could have jumped a distance of 2 m · 4 = 8 m.

- As mentioned, an angle of 45° will always result in the maximum range for a certain velocity. In the case of the dolphin, choosing the optimal angle increased the range by over 50 %.

- If the gravitational acceleration is doubled, the range is halved. This is an inverse proportional relationship. On the moon, where g has only one-sixth of earth's value, the dolphin could have jumped 2 m · 6 = 12 m.

Another quantity that might be of interest in this case is the maximum height reached H (in m). You can compute it with this somewhat complicated looking equation:

$$H = 0.5 \cdot (v \cdot \sin(\theta))^2 / g$$

For our dolphin, we get a maximum height of H = 1.4 m for θ = 70° and H = 0.8 m for θ = 45°. As you can see, getting as far as possible does not always mean getting as high as possible.

- **Impact Velocity:**

We will stick to gravity for yet another great formula. For many applications it is necessary (or just interesting) to know the speed at which a dropped object impacts the ground. To calculate it, we need two quantities: the drop height h (in m) and the gravitational acceleration g (in m/s^2). Using the conservation of energy, one can derive this formula for the impact velocity v (in m/s):

v = sq root (2 · g · h)

Simple, isn't it? But note that air resistance was neglected, which means that with an atmosphere present, the computed value will only be an approximation. The formula produces the most accurate values for heavy objects dropped from lower heights. In the section "Energy Conservation" you will see how this formula is derived.

A crane accidentally drops a heavy girder from a height of 20 m. At what speed will it impact?

v = sq root (2 · 9.81 m/s^2 · 20 m)

v ≈ 20 m/s = 72 km/h ≈ 45 mph

In this case the approximation should be very well as the girder is heavy and not influenced significantly by air resistance over such a short distance.

Let's see how changing the input quantities impacts the impact velocity:

- If you quadruple the drop height (or the gravitational acceleration), the impact velocity doubles. So dropping the girder from 20 m · 4 = 80 m will cause it to land with 45 mph · 2 = 90 mph.

Once the drop height gets too big and air resistance becomes a determining factor, the formula doesn't work anymore. Luckily there's another formula for just this case. After a while of free fall, any object will reach and maintain a terminal velocity. To calculate it, we need a lot of inputs.

The necessary quantities are: the mass of the object (in kg), the gravitational acceleration (in m/s^2), the density of air D (in kg/m^3), the projected area of the object A (in m^2) and the drag coefficient c (dimensionless). The latter two quantities need some explaining.

The projected area is the largest cross-section in the direction of fall. You can think of it as the shadow of the object on the ground when the sun's rays hit the ground at a ninety degree angle. For example, if the falling object is a sphere, the projected area will be a circle with the same radius.

The drag coefficient is a dimensionless number that depends in a very complex way on the geometry of the object. There's no simple way to compute it, usually it is determined in a wind tunnel. However, you can find the drag coefficients for common shapes in tables.

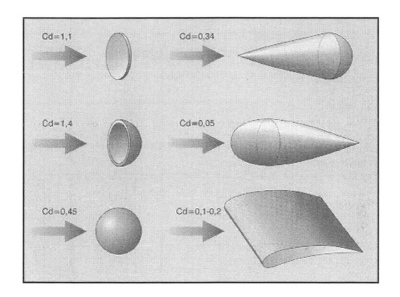

Now that we know all the inputs, let's look at the formula for the terminal velocity v (in m/s). It will be valid for objects dropped from such a great heights that they manage to reach this limiting value, which is a result of the air resistance canceling out gravity.

v = sq root (2 · m · g / (c · D · A))

Let's do an example.

Skydivers are in free fall after leaving the plane, but soon reach a terminal velocity. We will set the mass to m = 75 kg, g = 9.81 (as usual) and D = 1.2 kg/m³. In a head-first position the skydiver has a drag coefficient of c = 0.8 and a projected area A = 0.3 m². What is the terminal velocity of the skydiver?

v = sq root (2 · 75 · 9.81 / (0.8 · 1.2 · 0.3))

v ≈ 70 m/s ≈ 260 km/h ≈ 160 mph

According to some reports, there have already been injuries due to coins being dropped from tall buildings such as the Empire State. Is that possible? How fast would a coin dropped from such heights hit the ground (or people walking on it)?

Let's collect the necessary inputs. If the coin falls flat, it has a drag coefficient of c = 1.1 (see image). A dime has a mass of m = 0.002 kg and a radius of about 8 mm = 0.008 m, which, as you will learn in the section "Going in Circles", corresponds to an area of A = 0.0002 m². The air density near the ground is D = 1.25 kg/m³. Now let's apply the formula to see what the coin's terminal velocity will be:

v = sq root (2 · 0.002 · 9.81 / (1.1 · 1.25 · 0.0002))

v ≈ 12 m/s ≈ 43 km/h ≈ 27 mph

This is for example much less than an airsoft pellet, which at close distance can impact with 100 m/s. So at this speed the coin could only cause injuries if it fell directly on a person's eyes. Otherwise it would hurt a little and that's about it.

Again, let's take a look how changing the inputs varies the terminal velocity. Two bullet points will be sufficient here:

- If you quadruple the mass (or the gravitational acceleration), the terminal velocity doubles. So a very heavy skydiver or a regular skydiver on a massive planet would fall much faster.

- If you quadruple the drag coefficient (or the density or the projected area), the terminal velocity halves. This is why parachutes work. They have a higher drag coefficient and larger area, thus effectively reducing the terminal velocity.

For now, let's move away from gravity and its consequences. But we will surely return to it in later sections.

- **Braking distance:**

If something unexpected happens while driving, hitting the brakes should be your first impulse. How long it then takes for you to come to a complete halt depends on many things. Here are the quantities you need to compute the braking distance.

An important factor is obviously the current speed v (in m/s). Aside from that, we also need the reaction time t (in s) and the deceleration (in m/s²). The latter will depend mainly on how strongly you hit the brakes and the conditions of the road. If you've ever had the misfortune of needing to stop a car on snow or ice, you certainly can confirm this. The formula for the braking distance d (in m) is:

$$d = v \cdot t + v^2 / (2 \cdot a)$$

Before going to the example, let's first look at typical values for the inputs. For an alert and sober driver the reaction time is t = 1 s. When the driver is intoxicated or writing text messages, this can increase to t = 2 s. As for the deceleration, typical values are a = 8 m/s² on dry asphalt, a = 6 m/s² on wet asphalt, a = 2.5 m/s² on snow and a = 1 m/s² on ice, all in the case of full braking.

A sober driver (t = 1 s) hits the brakes at v = 75 mph ≈ 34 m/s on dry asphalt (a = 8 m/s²). What is his braking distance?

d = 34 · 1 + 34² / 16 ≈ 105 m ≈ 350 ft

How does this compare to a drunk driver (t = 2 s) under the same conditions? We apply the formula again:

d = 34 · 2 + 34² / 16 ≈ 140 m ≈ 460 ft

The drunk driver's braking distance is thus 35 m (or about 8 car lengths) longer. It goes without saying that this significant increase can decide between life and death.

Let's go back to the sober driver and see what the 105 m braking distance turns into when the road is icy:

d = 34 · 1 + 34² / 2 ≈ 610 m ≈ 2000 ft

Surprised? Recalling last winter, I'm not. This value is why no informed, sane person would ever consider going more than 30 mph or so when the road is fully covered in ice. Even at 30 mph the braking distance is still circa 105 m.

As for the dependencies, it is important to note that the braking distance increases with the square of the velocity rather than being proportional. This means that if you double the speed, the braking distance will increase (approximately) fourfold.

- **Centrifugal Force:**

When you drive your car through a curve, you notice that there's a force pushing you to the side. It is caused by your inertia. Your body wants to move forwards in a straight line, but the car disagrees and forces you to turn. What you experience is the centrifugal force.

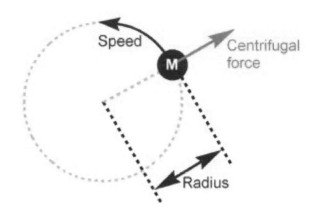

The centrifugal force acts on a body whenever it changes direction. It is not present in motion along a straight line. When it acts, it always does so perpendicular to the direction of travel and away from the center of the curve. So this force tries to "throw you out" of a curve.

Let's turn to the formula. It will work whenever we have circular motion (most curves are parts of circles, so it works there as well). As inputs we need the mass of the moving object m (in kg), its velocity v (in m/s) and the radius of the curve r (in m). Given these quantities, we can easily compute the magnitude of the centrifugal force:

$$F = m \cdot v^2 / r$$

Let's turn to an example.

Assume the average adult of mass m = 75 kg drives with v = 35 m/s through a curve of radius r = 400 m. He will experience this centrifugal force:

$$F = 75 \ kg \cdot (35 \ m/s)^2 / 400 \ m \approx 230 \ N$$

Is that noticeable? Well, let's compare it to the gravitational force. We calculated that the average adult experiences a gravitational pull of 785 N, the centrifugal force in this case is a little less than one third of that. Very much noticeable!

Let's turn an analysis of how changing the inputs impacts the magnitude of the centrifugal force:

- If we double the mass of the moving object, the centrifugal force doubles as well. So a 160 kg person would experience a force of 230 N $\cdot$ 2 = 460 N.

- If we double the velocity of the object, the centrifugal force increases fourfold. Because of this strong variation, even a little too fast can throw your car and you out of the curve. So choose your speed carefully. The average adult driving with with 35 m/s $\cdot$ 2 = 70 m/s through the curve from our example, would be pulled with 4 $\cdot$ 230 N = 920 N sideways, which is more than the gravitational force.

- If we double the radius of the curve, thus making the curve wider, the centrifugal force halves. An increase in radius to 400 m · 2 = 800 m would decrease the force experienced to 230 N : 2 = 115 N.

Combing the formula for the centrifugal force with the law of gravitation can lead to very powerful results, as can be seen in the next section.

- **Satellites:**

To understand this part, you need to read the sections "Gravity" and "Centrifugal Force", because here we will combine the two. Why don't satellites fall to the ground? After all, there's gravity pulling them to the earth's surface at all times. Still, they stay up there, orbiting the planet.

The fact that they keep orbiting despite gravity trying to ground them shows that there must be a second force canceling gravity out. As you might have guessed by the introduction, this is the centrifugal force. It acts away from the center of the circular motion, so in the exact opposite direction of the gravitational force.

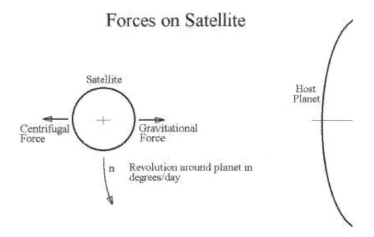

Forces on Satellite

For the two forces to cancel each other out, they not only need to have opposite directions but also the same magnitude. So this equation must hold true:

$G \cdot m \cdot M / r^2 = m \cdot v^2 / r$

On the left side is the expression for the gravitational, on the right side that of the centrifugal force. As for the quantities: m is the mass of the satellite, M the mass of earth, r the distance between the satellite and earth (or to be more specific: from the satellite to the center of earth) and v the velocity of the orbiting satellite.

What's so great about this equation is that we can deduce a neat formula that shows how the velocity v of a satellite depends on its distance r to earth's center:

v = sq root (G · M / r)

Did you notice what happened to the satellite's mass? During the process of solving for v it simply vanished. This is nice because that's one less variable to know. So the mass of a satellite does not impact the speed it will have in orbit at all.

A satellite is orbiting at a distance r = 30,000,000 m from earth's center. What is its speed? For that we also need to know the mass of earth, which you can find in the section "Gravity". Inputting the values leads to:

v = square root (G · 5.97 · 10²⁴ kg / 30,000,000 m)

v ≈ 3640 m/s ≈ 13,100 km/h ≈ 8130 mph

Let's take a look at out how altering the input quantities changes the orbital speed.

- If you increase the planet's mass fourfold, the orbital speed doubles. So at the given distance, orbiting a planet four times the mass of earth would result in a speed of about 3640 m/s · 2 = 7280 m/s.

- If you increase the satellite's distance to the planet fourfold, its velocity halves. So bringing the satellite up to a distance of 30 Mm · 2 = 60 Mm would reduce its speed to 3640 m/s : 2 = 1820 m/s.

We will take a look at more examples, but before we can do that, we need to do some preparation. Let's deduce a formula for the rotation period of the satellite. If the satellite is in orbit at a distance r from earth's center, it must travel a distance of:

$$d = 2 \cdot \pi \cdot r$$

to complete one revolution. More on this formula in the section "Going in Circles". Remember that to compute how long it takes an object to travel a certain distance, we divide the distance by the velocity. For example, if you need to travel d = 400 miles and your average speed is v = 80 mph, then the trip will take you: T = d / v = 5 hours. Similarly, it will take the satellite this time T (in s) to complete one revolution:

$$T = 2 \cdot \pi \cdot r / v$$

Since we already derived a formula for the speed of the satellite, let's insert it here. This leads to:

$$\mathbf{T = 2 \cdot \pi \cdot sq\ root\ (r^3 / G \cdot M)}$$

This formula allows us to calculate a very special orbit as you will see in the second example. Before we do that, let's revisit the satellite from the previous example.

We already determined that a satellite orbiting at a distance
of r = 30,000,000 m from earth's center will move at a
speed of v = 3640 m/s. So instead of blindly applying the
formula, we'll just compute the distance it travels during one
revolution and compute the rotation period from that.

$d = 2 \cdot \pi \cdot 30,000,000 \; m$

$d \approx 188,400,000 \; m$

As mentioned, the rotation period is just the distance divided
by the speed. Thus we get:

$T = d / v \approx 51,760 \; s \approx 14.4 \; h$

In the time it takes the earth to do six revolutions (24 h $\cdot$ 6 =
114 h), the satellite will complete ten (14.4 h $\cdot$ 10 = 114 h).

As we increase the distance of the satellite to earth, the
rotation period increases as well. This means that at a
certain distance, the satellite will have the same rotation
period as earth (24 h) and thus will always hover above the
same point. So for an observer on earth, the satellite seems
to be stationary. Hence the name geostationary orbit for this
special orbit. At what distance from the center of earth is the
geostationary orbit located?

24 hours correspond to 86,400 seconds. Using the formula
for the rotation period and this value we can easily set up an
equation for the radius of the geostationary orbit (now you
know why we needed to do some preparation):

$86,400 = 2 \cdot \pi \cdot sq\ root\ (r^3 / G \cdot M)$

Dividing by 2 · π and squaring leads to:

$(43,200 / \pi)^2 = r^3 / G \cdot M$

Now we multiply both sides by G · M and apply the third root, resulting in this value for the radius of the geostationary orbit:

$r \approx 42,200\ km$

Subtracting earth's radius of 6400 km, this means that the satellite is at an altitude of 35.800 km above earth's surface. Just to put that into perspective: commercial planes fly at heights of 10 to 12 km.

In 1945 the science-fiction author Arthur C. Clarke already proposed to place satellites at this altitude to make world-wide radio communication possible. In 1963 Syncom 2 became the first operational satellite in geostationary orbit. Today more than 250 satellites are in place there.

During the course of this book we will see more examples of how we can combine two forces to produce very useful results. The next chapter is such an example and we stick to the two forces we just talked about.

- **Roller Coaster Loops:**

Roller Coaster Loops are the highlight of any visit to an amusement park. They are the ultimate thrill, though I have to admit that personally I prefer to stay away from them. Let's just say you must have the right stomach for this experience.

Again we should ask ourselves why the train doesn't just fall down during the loop and again the answer will be: the centrifugal force cancels out gravity. This time we can use the simplified formula for the gravitational force as all of this happens at the surface of earth.

We denote the mass of the train by m (in kg), the velocity at the top of the loop by v (in m/s) and the radius of the loop by r (in m). To derive the minimum speed required to complete the loop successfully, we'll set the formula for the gravitational force equal to that of the centrifugal force:

$m \cdot g = m \cdot v^2 / r$

v = sq root (r · g)

Just like in the case of the satellites, the mass does not turn out to be an influencing factor. A more massive train will have a higher gravitational pull as well as a higher centrifugal push, so the ratio of the two remains the same. This is great because this way, we don't have to worry about how the equilibrium of the forces is affected by varying numbers of people in the train.

As you can see, the determining factor here is the radius of the loop. If you quadruple it, the required speed doubles. Of course the gravitational acceleration is in there as well, but usually we don't care that much for roller coasters on other planets or moons, so we can regard it as a constant.

Another quantity that is of interest here is the required entry and exit speed. Note that for the computation of the force equilibrium we just needed the speed at the top of the loop. But naturally this speed is a result of how fast we enter the loop. We can derive the formula for the entry and exit speed u (in m/s) using the energy conservation law, which will be featured in a later section of this book.

u = sq root (5 · r · g) ≈ 2.24 * v

This relation holds true assuming that ground friction and air resistance can be neglected (which is usually a good approximation for roller coaster loops).

The largest roller coaster loop can be found at Six Flags Magic Mountain in Valencia, California. Its radius is roughly r = 22 m. How fast does a train need to be at the top of the loop to not drop down? What is the minimum entry speed for the loop?

Let's turn to the first question:

v = sq root (22 m · 9.81 m/s²)

v ≈ 15 m/s ≈ 53 km/h ≈ 33 mph

At the top of the loop it needs move with at least 33 mph. As for the minimum entry speed, according to the second formula this is just the velocity at the top times 2.24:

u ≈ 33.5 m/s ≈ 120.5 km/h ≈ 75 mph

Often the entry speed is provided by a preceding sharp drop in height (which by itself is quite the exciting experience). What height difference do we need so that the drop provides the necessary entry velocity? In the section "Impact Speed" you learned that an object being dropped and falling freely over a height h gains this speed v:

v = sq root(2 · g · h)

Interestingly enough, this formula also works for our roller coaster train as long as we can neglect frictional forces. So to calculate the necessary drop in height h (in m) to gain the entry speed u (in m/s), we just rearrange the formula above:

h = u² / (2 · g) = 2.5 · r

After inserting the second formula of this section, the one for the entry speed, into this formula, we can conclude that the required drop in height is always 2.5 times the radius.

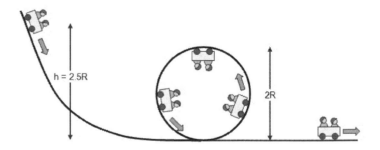

Let's revisit the largest roller coaster loop with its radius of r = 22 m. If a train, driven only by gravity, should successfully complete the loop, it must drop by at least h = 55 m ≈ 180 ft before entering the loop.

Of course in reality the situation is a little more complex than this. Ground friction and air resistance must be taken into consideration and the loops are often not exact circles but rather have a clothoidal form. But as a first approximation, the above formulas do a fantastic job.

- **Lift:**

Just like the satellites we just talked about, planes resist the pull of gravitation. However, they do it by very different means. So let's look at why flying works. When an airplane flies, some of the air is forced to go over and some of it to go under the wing. Because of the geometry of the wing, the air that flows over the wing goes faster than the air below it. This is a very crucial point because, according to the Bernoulli principle, the faster the air flows, the smaller the pressure is.

So splitting the air using the wing causes a pressure difference. Above the wing, the air moves fast and thus the pressure is low, while below the wing, we have the opposite situation. This pressure difference pushes the wing, and with it the plane, upwards. The resulting force is called lift and there's a simple formula to calculate it.

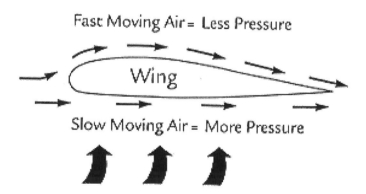

Here are the quantities we need for understanding the formula: the velocity v (in m/s) of the plane relative to the air, the area A (in m²) of the wing, the density D (in kg/m³)

of the air surrounding the plane and a less known quantity called the coefficient of lift c (dimensionless). The coefficient of lift depends on the specific geometry of the wing and the angle between the wing chord and the streaming air, which is known as the angle of attack.

Given these quantities, calculating the lift (in N) couldn't be easier. We just need to plug them into this formula:

$$L = 0.5 \cdot c \cdot D \cdot A \cdot v^2$$

A Boeing 747, also called Jumbo Jet, has a lift coefficient of c = 0.3 when the angle of attack is zero. We want to calculate the lift at cruising altitude. In these heights the air density is approximately D = 0.3 kg/m³. The wing area of a Jumbo Jet is A = 511 m² and the cruising speed v = 305 m/s. This is all we need to compute the lift:

$$L = 0.5 \cdot 0.3 \cdot 0.3 \ kg/m^3 \cdot 511 \ m^2 \cdot (305 \ m/s)^2$$

$$L \approx 2{,}150{,}000 \ N$$

So how does this compare to the gravitational force on (in short: weight of) the Jumbo Jet? Well, we don't even need to compute that. In order to stay at cruising altitude, the lift must cancel out the gravitational force. So lift and weight must be the same at level flight.

How the the parameters in the formula impact lift?

- If you double the coefficient of lift, the lift doubles as well. So if the Jumbo Jet increases its lift coefficient to 0.3 · 2 = 0.6 (for example by increasing the angle of attack), the lift increases to 2150 kN · 2 = 4300 kN. Doubling the density of air D and the wing area A has the same impact on lift.

- If you double the velocity of the plane, the lift increases fourfold. For the Jumbo jet this means that at a velocity of 305 m/s · 2 = 610 m/s, it would experience a lift of 2150 kN · 4 = 8600 kN (and probably break apart very quickly as the material couldn't handle such velocities).

There's another formula that is helpful in calculating lift. The air density is not a constant, it varies with height. The higher you go, the less dense the air becomes. The formula below can be used to approximate the air density D (in kg/m^3) at a certain height h (in m):

D = 1.25 · exp(-0.0001 · h)

According to this formula, the air density at sea level (h = 0 m) and at the top of the Mount Everest (h = 8850 m) is respectively:

D = 1.25 · exp(-0.0001 · 0 m) = 1.25 kg/m³

D = 1.25 · exp(-0.0001 · 8850 m) ≈ 0.52 kg/m³

How does this variation in density impact lift? Remember that as the air density shrinks, so does lift. Thus, when you increase altitude, lift decreases. Or in other words: as you go up, you need to go faster to maintain a certain amount of lift (which is just what planes do when they climb).

Another consequence of this is that the necessary take-off speed is bigger for high-altitude airports. The fact that at greater heights the plane's engines produce less thrust (as well as less reverse thrust) makes taking off and landing at such airports even more challenging.

- **Airplane Speed:**

It was already mentioned in the previous section that in order to achieve level flight, you need the lift to cancel out the gravitational pull exactly. We can formulate that mathematically by setting lift equal to weight.

$$0.5 \cdot c \cdot D \cdot A \cdot v^2 = m \cdot g$$

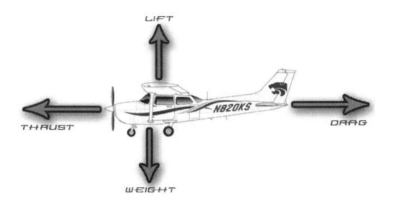

Note that I used the simplified formula for the gravitational force, which you might find odd since I stated that it only holds true at earth's surface (on which we clearly are not in this case). But remember the dimensions we are talking about here. Earth's radius is about 6400 km, whereas planes usually fly in altitudes 12 km or below. So for most practical purposes, 12 km altitude is still at the surface of earth.

What's nice about the above combination of lift and weight is that we can compute an airplane's equilibrium speed from it by solving for v. The result looks like this:

v = sq root (2 · m · g / (c · D · A))

A loaded Cessna 152 has a mass of about m = 700 kg, lift coefficient c = 1.2 and wing area A = 15 m². What is its equilibrium speed at very low altitudes? We will use the air density at sea level D = 1.25 kg/m³ and g = 9.81 m/s². Inputting this into the above formula leads to:

v = sq. root (2 · 700 · 9.81 / (1.2 · 1.25 · 15))

v ≈ 25 m/s = 90 km/h ≈ 56 mph

Let's look at how varying the inputs will alter the equilibrium speed of an airplane:

- If you quadruple the plane's mass, the equilibrium speed doubles. A plane four times the mass of the Cessna 152 (with all other parameters unchanged), would need to fly at 56 mph · 2 = 112 mph to maintain its altitude.

- If you quadruple the lift coefficient (or density or wing area), the equilibrium speed halves. So if we were able to find a wing geometry that would result in the lift coefficient being quadrupled, the Cessna 152 could fly as slow as 56 mph / 2 = 28 mph and still remain in level flight.

This section showed once more that we can get powerful results by combining formulas. The concept of equilibrium of forces worked for satellites as well as for airplanes. Whenever an object maintains height, stands still or moves in a straight-line, there must be an equilibrium of forces causing that.

- **Momentum:**

Now we get to a very fundamental physical quantity, linear momentum. Its definition is very simple. An object with mass m (in kg) moving at the velocity v (in m/s) has the momentum:

$p = m \cdot v$

So a massive object moving very fast has a lot of momentum, while a light object moving very slow has only little. If an object doesn't move at all, its momentum is zero. So why should we care about that? It seems rather artificial to define such a quantity. But here's the beautiful part: in any system of objects, total momentum is conserved. So if one object loses momentum, the other objects have to gain exactly this amount. In mathematical terms:

p = constant

And this is our great formula. One cannot overstate the importance of it. It is as important as the conservation of energy and without it, we would still be in the dark about many aspects of nature. It works for a cluster of stars as well as for two billiard balls. As far as we know, this law true in all of the universe. It is how the universe works.

But for our example, let's not get carried away.

Recoil is a consequence of the conservation of momentum. Before a shot is fired, both the gun and the bullet are at rest. So the total momentum is zero. When the bullet is fired, it gains momentum in a certain direction. In order for momentum to be conserved, the gun must gain the same amount of momentum in the opposite direction.

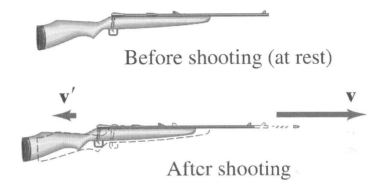

Before shooting (at rest)

After shooting

A typical 9 mm bullet has a mass of m = 0.012 kg and is launched at a velocity of about v = 450 m/s (which is more than the speed of sound). In the process of firing, it has gained the momentum:

$p = 0.012 \ kg \cdot 450 \ m/s = 5.4 \ kg \ m/s$

How does this affect a m' = 3 kg rifle? Since the momentum must be conserved, this equation must hold true:

$5.4 \ kg \ m/s = 3 \ kg \cdot v'$

with v' being the velocity at which the rifle is thrown back. Solving this for v' results in:

$v' \approx 1.8 \ m/s \approx 6.5 \ km/h \approx 4 \ mph$

Another classic application of the conservation of momentum is the propulsion of rockets using gases. Imagine a rocket at rest in space. To gain velocity, it expels m =

3000 kg of hot gas at a speed of v = 4200 m/s. The rocket weighs m' = 10,000 kg. How fast will it go after the engine is shut down?

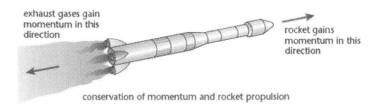

exhaust gases gain momentum in this direction

rocket gains momentum in this direction

conservation of momentum and rocket propulsion

The momentum of the exhaust gas is:

p = 3000 kg · 4200 m/s = 12,600,000 kg m/s

In order for the momentum to be conserved, the rocket must gain the same amount of momentum in the opposite direction:

12,600,000 kg m/s = 10,000 kg · v'

This leads to:

v' = 1260 m/s ≈ 4540 km/h ≈ 2810 mph

Note that it since there's no air resistance or gravity involved, it does not matter at which rate the rocket burns the gas. It could burn 30 kg/s for 100 seconds or 3 kg/s for 1000 seconds, the final velocity would be the same.

In the next sections we'll deal with yet another conservation law. Such conservation laws are very useful in doing calculations and understanding how nature works, so you should make sure to know them by heart.

- **Energy:**

This section is meant as preparation for the next section, in which we will talk about conservation of energy. To do that, we need to know some common types of energy. We will focus on types of energy that are used in computing motion. All energy is measured in Joules (J) or units derived from that.

One energy form we will repeatedly need is kinetic energy E(kin). It is the energy an object possesses due to its speed and thus also the minimum energy needed to bring an object to a certain velocity. It depends only on two inputs: the mass of the object m (in kg) and its velocity (in m/s).

$$E(kin) = 0.5 \cdot m \cdot v^2$$

Note that there's a quadratic dependence on speed meaning that if you double speed, the kinetic energy quadruples.

How do the kinetic energies of small and large airplanes compare? A loaded Cessna 152 has a mass of m = 700 kg and cruises at a speed of about v = 53 m/s. Its kinetic energy is:

$$E(kin) = 0.5 \cdot 700 \ kg \cdot (53 \ m/s)^2$$

$$E(kin) \approx 983,000 \ J$$

So a small airplane at cruising altitude has a kinetic energy in the order of one million joules. What about a fully loaded Jumbo Jet? The newest models have a mass of m = 400,000 kg and cruise at v = 255 m/s. This translates into:

E(kin) = 0.5 · 400,000 kg · (255 m/s)²

E(kin) ≈ 13,000,000,000 J

So here the kinetic energy is about 13 billion joules. With the kinetic energy required to bring one Boeing 747 to its cruising altitude you could you could do the same for 13,000 Cessna 152. That really puts things into perspective.

Another form of energy objects can possess and that is relevant to motion is the potential energy E(pot). It is the energy an object has due to its location in a gravitational field. If a body is at a very high altitude, there's the potential for the release of a large amount of energy by it dropping. For objects close to the surface, the only quantities involved are the mass of the object m (in kg), the gravitational acceleration g (in m/s²) and its height h (in m).

E(pot) = m · g · h

All relationships here are linear. If you double the mass (or gravitational acceleration or height), the potential energy doubles as well. Let's look at an example.

Let's stick with the Cessna and the Jumbo Jet. They have respective cruising altitudes of h = 2200 m and h = 11,000 m. What are their potential energies in cruising mode? For the gravitational acceleration we'll use g = 9.81 m/s² as always. First let's look at the Cessna:

E(pot) = 700 kg · 9.81 m/s² · 2200 m

$E(pot) \approx 15,100,000 \, J$

On to the Jumbo Jet:

$E(pot) = 400,000 \, kg \cdot 9.81 \, m/s^2 \cdot 11,000 \, m$

$E(pot) \approx 43,200,000,000 \, J$

We will include one more energy form relevant to motion before moving on to the "real deal", that is, energy conservation. This energy form is frictional energy E(fric). The name says it all: it is the amount of energy we need to provide to overcome frictional forces. To keep things simple, we restrict ourselves to ground friction.

The quantities involved here are: the mass m (in kg) of the object that is in motion, the gravitational acceleration g (in m/s^2) and the distance d (in m) the object travels. Aside from that we also need the coefficient of friction μ (dimensionless), which depends on the material of the ground, the material of the object and the nature of the contact between the two.

$E(fric) = \mu \cdot m \cdot g \cdot d$

Again all relationships are linear, if you double one of the inputs, the frictional energy doubles as well. Let's turn to an example.

We want to (or rather have to) displace a m = 100 kg concrete block by a distance of d = 10 m on wood ground. The coefficient of friction of concrete on wood is $\mu \approx 0.6$. How much energy do we need to overcome friction?

$E(fric) = 0.6 \cdot 100 \, kg \cdot 9.81 \, m/s^2 \cdot 10 \, m$

$E(fric) \approx 5900 \, J$

So if we were able to provide a power of 200 watt = 200 J/s this would take us about 30 seconds.

Instead of just displacing the m = 100 kg concrete block by pushing it over the ground, we'll lift it on a small cart and push the cart. The friction coefficient is reduced to $\mu \approx 0.03$. However, now we also need to provide potential energy to lift the block to a height of h = 0.25 m. Is this approach smarter in terms of energy?

The frictional energy we need to provide is:

$E(fric) = 0.03 \cdot 100 \, kg \cdot 9.81 \, m/s^2 \cdot 10 \, m$

$E(fric) \approx 295 \, J$

And this is the potential energy we need to lift the concrete block on the cart:

$E(pot) = 100 \, kg \cdot 9.81 \, m/s^2 \cdot 0.25 \, m$

$E(pot) \approx 245 \, J$

We do not include any energy for getting the block off the cart. This work can be done by gravity alone (not elegantly, but still). So to displace the concrete block by ten meters using the cart we had to provide in total 295 J + 245 J = 540 J, much less than the 5900 J we needed to push the block over the ground by brute force.

- **Energy Conservation:**

Now that we know some common energy types and how to compute them, we are ready to take a look at one of the most (if not the most) fundamental principal of physics. And you don't need to be a man or woman of many words to state the energy conservation law: the total amount of energy in any system is constant. Or in mathematical terms:

E = const.

That's it, pure and simple. There is no if, no when, no but. All processes that have been observed to this date, whether under a electron microscope or in the depth of space, have fully obeyed this law. As far as we know, it holds true on any scale and in any part of the universe. The application of the conservation of energy has led to many great formulas and discoveries. We will only take a quick peek into this rich field but rest assured you could fill entire volumes with it.

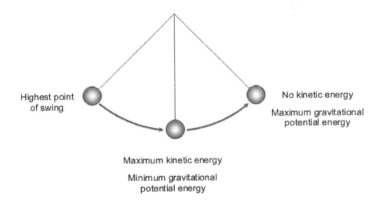

Highest point of swing

No kinetic energy

Maximum gravitational potential energy

Maximum kinetic energy

Minimum gravitational potential energy

As a first application, let's take a look at free fall. Note that the following deliberations also work just fine for pendulum swings or almost frictionless motion on the ground over different heights. From the perspective of energy conservation, it doesn't make much of a difference.

When a body is at a great height, it possesses a lot of potential energy. As it drops, it loses potential energy. At the same time it gains speed, which means that the kinetic energy increases. If there are no other forces except gravity involved (no friction), then the sum of the potential and kinetic energy must be constant:

E(kin) + E(pot) = const.

This allows us to compute the velocity at any given height. Note that initially all the energy is in form of potential energy. When the object reaches the ground, all of the initial potential energy has been transformed into kinetic energy. So if we only care about the impact velocity, we can state the energy conservation in this form:

E(kin, ground) = E(pot, initial)

If we denote the initial height with h (in m) and the impact velocity with v (in m/s), we get this equation:

$0.5 \cdot m \cdot v^2 = m \cdot g \cdot h$

From that we can easily deduce the impact speed:

v = sq root (2 · g · h)

Does this formula look familiar to you? If you read the section "Impact Speed" it should, as we already took a look at it there. Now you know where this great formula comes from. It is simply the consequence of the energy conservation law. So let's go right to the next application.

When you stop applying pressure on the gas pedal in your car, the car will slowly but surely roll to a stop. As it does so, it loses kinetic energy. Where does it go? Assuming the road is horizontal, there will be no change in potential energy. So it must transform into frictional energy (again neglecting air resistance). During the process of rolling to a halt, the sum of kinetic and frictional energy must remain the same in order for the conservation law to hold true:

E(fric) + E(kin) = const.

This allows us to compute the velocity after rolling a certain distance. Again we are mainly interested in the final state, when all the initial kinetic energy has fully transformed into frictional energy. In mathematical terms:

E(fric, final) = E(kin, initial)

Denoting the initial speed by v (in m/s) and the distance over which the car rolled to a halt by d (in m), we get:

$\mu \cdot m \cdot g \cdot d = 0.5 \cdot m \cdot v^2$

So the distance over which the car rolls out is:

$d = 0.5 \cdot v^2 / (\mu \cdot g)$

Did you notice what happened to the mass? Again it simply vanished. The mass of the car has no impact on the distance over which it rolls out. Also noteworthy is that the dependence on initial speed is quadratic, meaning that if you double the speed of the car, the distance over which it rolls out increases fourfold. Isn't it amazing what we can deduce by simply applying the conservation law?

The coefficient of friction for a car tire rolling on asphalt is about μ ≈ 0.015. Over what distance does a car driving at 30 mph = 13.5 m/s roll out? We already derived the necessary formula, so all we need to do is to plug in the values.

$d = 0.5 \cdot 13.5^2 / (0.015 \cdot 9.81) \approx 620\,m \approx 2000\,ft$

Since we neglected air resistance, the actual value is going to be a bit smaller than that. As the speed grows, the influence of air resistance gets higher and must be included. So the formula we derived from the energy conservation has its limits, but this is not the conservation law's fault, it is ours for leaving out other factors at play.

A hope this section helped you to appreciate the meaning and usefulness of the energy conservation law in physics. If you want to be serious about physics, be sure to learn as many energy types as possible by heart and how to combine them to derive new formulas.

- **Heat:**

A long time ago, in my teen years, this was the formula that got me hooked on physics. Why? I can't say for sure. I guess I was very surprised that you could calculate something like this so easily. So with some nostalgia, I present another great formula from the field of physics. It will be a continuation of and a last section on energy.

To heat something, you need a certain amount of energy E (in J). How much exactly? To compute this we require three inputs: the mass m (in kg) of the object we want to heat, the temperature difference T (in °C) between initial and final state and the so called specific heat c (in J per kg °C) of the material that is heated. The relationship is quite simple:

$$E = c \cdot m \cdot T$$

If you double any of the input quantities, the energy required for heating will double as well. A very helpful addition to problems involving heating is this formula:

$$E = P \cdot t$$

with P (in watt = W = J/s) being the power of the device that delivers heat and t (in s) the duration of the heat delivery.

The specific heat of water is c = 4200 J per kg °C. How much energy do you need to heat m = 1 kg of water from room temperature (20 °C) to its boiling point (100 °C)? Note that the temperature difference between initial and final state is T = 80 °C. So we have all the quantities we need.

$$E = 4200 \cdot 1 \cdot 80 = 336,000 \, J$$

Additional question: How long will it take a water heater with an output of 2000 W to accomplish this? Let's set up an equation for this using the second formula:

$336,000 = 2000 \cdot t$

$t \approx 168 \text{ s} \approx 3 \text{ minutes}$

We put m = 1 kg of water (c = 4200 J per kg °C) in one container and m = 1 kg of sand (c = 290 J per kg °C) in another next to it. This will serve as an artificial beach. Using a heater we add 10,000 J of heat to each container. By what temperature will the water and the sand be raised?

Let's turn to the water. From the given data and the great formula we can set up this equation:

$10,000 = 4200 \cdot 1 \cdot T$

$T \approx 2.4 \text{ °C}$

So the water temperature will be raised by 2.4 °C. What about the sand? It also receives 10,000 J.

$10,000 = 290 \cdot 1 \cdot T$

$T \approx 34.5 \text{ °C}$

So sand (or any ground in general) will heat up much stronger than water. In other words: the temperature of ground reacts quite strongly to changes in energy input while water is rather sluggish. This explains why the climate near oceans is milder than inland, that is, why the summers are less hot and the winters less cold. The water efficiently dampens the changes in temperature.

It also explains the land-sea-breeze phenomenon (seen in the image below). During the day, the sun's energy will cause the ground to be hotter than the water. The air above the ground rises, leading to cooler air flowing from the ocean to the land. At night, due to the lack of the sun's power, the situation reverses. The ground cools off quickly and now it's the air above the water that rises.

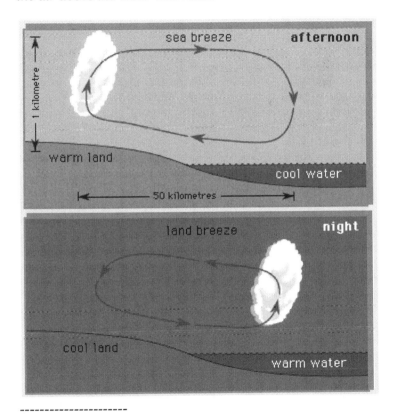

I hope this formula got you hooked as well. It's simple, useful and can explain quite a lot of physics at the same time. It doesn't get any better than this. Now it's time to leave the concept of energy and turn to other topics.

Part II: Mathematics

- **Trigonometry:**

In geometry you will commonly deal with right triangles and trying to compute them without the incredibly useful trigonometric formulas is just madness. They are the screwdrivers in the physicist's and mathematician's toolbox, you always need to have them with you or the simplest problems can quickly become unsolvable.

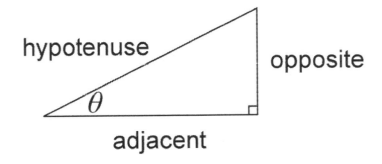

In the picture above you can see a right triangle, that is a triangle, that has one 90° angle. The side opposite the right angle is called **hypotenuse**. It is always the longest side in the triangle. Let's pick an angle other then the right angle and denote it by θ (in °). Unsurprisingly, the side adjacent to it is called the **adjacent** and the side opposite to it the **opposite**.

The great formula here will actually be three. They allow us to compute the entire triangle when two quantities are given. If we know two of the sides, we can calculate the third side and the angle, if we know the angle and one side, we can deduce the lengths of the remaining two sides. Here they are:

sin θ = opposite / hypotenuse

cos θ = adjacent / hypotenuse

tan θ = opposite / adjacent

Which one to use will be determined by which two quantities we are given. If we are given the adjacent and the hypotenuse, it obviously makes sense to start with the second formula. If we are given the angle θ and the hypotenuse, we can start with the first or second formula, depending on what we want to find out.

A plane takes off and remains at an angle of θ = 15° while climbing. What distance d has it flown when reaching its cruising altitude of h = 11 km?

Make a sketch or go back to the image above to visualize this situation. In this case the height is obviously the opposite of the given angle and the flown distance the hypotenuse. This means that we will need to use the sin-formula for our calculations.

sin θ = opposite / hypotenuse

sin 15° = 11 km / d

0.26 ≈ 11 km / d

In the last step we used a calculator to evaluate sin 15°. Make sure the calculator is set to "degrees" not "rad". Otherwise you will end up with an incorrect result. Now multiply both sides of the equation with d:

d · 0.26 ≈ 11 km

And then divide both sides by 0.26:

d ≈ 42 km

We want to determine the angle θ at which the sun's rays impact the ground. To do that, we place a box on the ground and measure its height and the length of its shadow. The respective values are h = 40 cm and l = 15 cm.

Again, visualize this situation using the image above. The height of the box is obviously the opposite, the length of the shadow the adjacent. So here we are required to apply the tan-formula.

tan θ = opposite / adjacent

tan θ = 40 / 15 ≈ 2.67

To deduce the angle from that, we need to make use of the calculator's inverse function. Enter the number and press "inverse" then "tan" (on some calculators you enter the number after pressing "inverse" and "tan"). To show that we are using the inverse function, we include the prefix "arc" in the equation.

θ = arctan(2.67) ≈ 69°

Using the box we just managed to show that at the moment the sun's rays impact the ground at a θ = 69° angle. If at the same time the shadow of a house is l = 8 m long, what's the height h of the house?

Going back to the image, we can see that again we're dealing with the opposite (the height) and the adjacent (shadow length) of the angle. So we'll stick to the tan-formula.

tan θ = opposite / adjacent

tan 69° = h / 8 m

Multiply both sides by 8:

h = 8 m · tan 69° ≈ 21 m

These were just a few of millions of possible applications for the trigonometric formulas. You can be sure that as you do mathematics, you will always want to (or at least need to) come back to them. Luckily, it's not rocket science, but rather a matter of making a sketch and identifying the sides correctly.

- **Going in Circles:**

For mathematicians the number π has an almost magical attraction. Most great names in mathematics have tried to find means to calculate it even more efficiently or spent time analyzing its nature at some point in their lives. In some ways this number is the border between the realm of the linear, straight-lined world humans have constructed and the non-linear, curved world that is nature. One of the greatest thing it does is enabling us to do calculations with circles and spherical objects.

A circle is a two-dimensional set of points all having a fixed distance to a center. This distance is called the radius r (in m). It is the only input we will need here. A sphere has the same definition as the circle with the exception that extends into the third dimension.

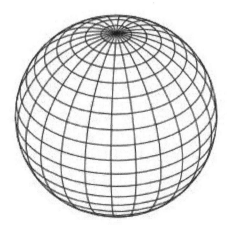

The great formulas featured in this section allow us to compute the circumference of a circle C (in m), the area of a circle A (in m^2), the surface area of a sphere S (in m^2) and the volume of a sphere V (in m^3). Doing geometry would be practically impossible without these.

$$C = 2 \cdot \pi \cdot r$$

$$A = \pi \cdot r^2$$

$$S = 4 \cdot \pi \cdot r^2$$

$$V = 4/3 \cdot \pi \cdot r^3$$

Here are some examples on how to apply them.

The radius of earth is approximately r = 6400 km. How far do you need to travel at the equator to go around earth once? This question requires us to calculate the circumference of the equatorial circle. Applying the formula we get:

C = 2 · π · 6400 km ≈ 40,200 km

In a plane traveling at 1000 km/h (which is the speed of a common passenger jet), this would take us:

t = 40,200 km / 1000 km/h ≈ 40 h

How long do you think this would take by foot? The normal walking speed is about 5 km/h, but since we need to rest and sleep, we will rather use an average of 3 km/h.

t = 40,200 km / 3 km/h ≈ 13,400 h ≈ 560 days

About 30 % of earth's surface is land. What is the total area of land on earth? Again we use the value r = 6400 km for the radius. According to the formulas presented here, the surface area of earth is:

$S = 4 \cdot \pi \cdot (6400 \ km)^2 \approx 515$ million km^2

So the total area of land on earth is $0.3 \cdot 515 \approx 155$ million km^2. A side note: circa half of this land is habitable for humans and since there are about 7 billion people on earth today, we can conclude that there is $0.022 \ km^2$ habitable land available per person. This corresponds to a square with 150 m $\approx$ 500 ft length and width.

FIFA rules state that a soccer ball must have a circumference of about 70 cm. What is the radius and volume of such a ball? First we set up an equation for the radius:

70 cm $= 2 \cdot \pi \cdot r$

Dividing by $2 \cdot \pi$ leads to:

$r = 70$ cm $/ (2 \cdot \pi) \approx 11$ cm

Now we can compute the volume:

$V = 4/3 \cdot \pi \cdot (11 \ cm)^3 \approx 5580 \ cm^3$

We could just go on with more and more examples and we wouldn't run out any time soon. Keep these formulas in mind, they are simple and enormously useful at the same time. You can apply them whenever there's a circle or sphere in sight (which is surprisingly often).

- **Quadratic Equations:**

I can hardly think of a formula that is more often used in mathematics than this one. It's quite long and looks rather intimidating, but still most people who do mathematics know it by heart. I'm talking about the formula to solve quadratic equations.

So let's first take a look at what quadratic equations are. In the most general case they consist of three terms and three real numbers a, b and c. Also included is the unknown x, the value of which we want to find out.

$$a \cdot x^2 + b \cdot x + c = 0$$

We always need the first term, so a is not allowed to be zero. But the other terms do not always show up, in which case the respective value for b or c is set to zero.

$$3 \cdot x^2 - 48 = 0$$

Here the values of the constants are: a = 3, b = 0 and c = -48. Note that the minus-sign is part of the constant. If we leave it out, we will arrive at an incorrect solution. Now let's turn to the great formula for this section. It spits out the solutions of a quadratic equation when we insert the values of constants.

$$x = (-b \pm \text{sq root} (b^2 - 4 \cdot a \cdot c)) / (2 \cdot a)$$

Granted, the equation looks horrible. And the plus-minus-sign is not making things easier. Why do we need it? A quadratic equation generally has two solutions. The first solution we get by using the plus-sign, the second by using the minus-sign. An example will show that the formula is not as bad as it looks. If you input the correct constants and evaluate the resulting numbers, nothing will go wrong.

We want to solve the equation:

$x^2 - 6 \cdot x + 8 = 0$

The values of the constants are: a = 1, b = -6 and c = 8. Now let's apply the formula to solve it:

$x = (6 \pm sq\ root\ ((-6)^2 - 4 \cdot 1 \cdot 8)) / (2 \cdot 1)$

Note that since b = -6 we set -b = 6.

$x = (6 \pm sq\ root\ (4)) / 2$

$x = (6 \pm 2) / 2$

Thus, the first solution is:

$x = (6 + 2) / 2 = 4$

The second solution is:

$x = (6 - 2) / 2 = 2$

So it wasn't as bad as you might have thought. But you noticed that we should be very careful in extracting the constants from the equation and inputting them into the formula. Using incorrect signs is the number one cause of frustration with such equations, make sure to avoid this.

When you apply the brakes of your car on dry asphalt, the braking distance d (in m) depends on the initial speed v (in m/s) as such:

$d = v + v^2 / 16$

For more information, check out the section "Braking Distance". We would like to know at what speed the braking distance becomes d = 50 m. Thus we get this equation:

$50 = v + v^2 / 16$

let's bring it to the general form of a quadratic equation and apply the great formula to solve it.

$v^2 / 16 + v - 50 = 0$

The constants are: a = 1/16, b = 1 and c = -50.

$v = (-1 \pm sq\ root\ (1^2 - 4 \cdot 1/16 \cdot (-50))) / (2 \cdot 1/16)$

$v = (-1 \pm sq\ root\ (13.5)) / 0.125$

$v \approx (-1 \pm 3.7) / 0.125$

The first solution is:

$v \approx (-1 + 3.7) / 0.125 \approx 22\ m/s \approx 79\ km/h \approx 50\ mph$

The second solution will be negative and thus of no relevance here. This is often the case when solving quadratic equations in physics problems and we can discard nonsensical solutions without worries.

If you are serious about mathematics, you must be able to solve quadratic equations. There's no way around it. The same goes for the type of equations we will look at in the next section. Again there's one "magical" formula that will allow us to arrive at correct solutions.

- **Logarithmic Identity:**

There are some formulas which are so useful that you couldn't picture yourself doing mathematics without them. This identity is one of them. Though often underrated and overlooked, it is what enables us to solve exponential equations. These equations arise naturally in a vast amount of situations: population growth, radioactivity, statistics, banking, ... In their general form, they look like this:

$$a^x = b$$

with a and b being known numbers and x the unknown number we want to find out. The key for solving such equations is provided by this simple yet powerful logarithmic identity:

$$\mathbf{ln(\ a^x\ ) = x \cdot ln(a)}$$

ln is short for natural logarithm, a function that can be found on any good calculator. Thanks to the identity, the unknown is not in the exponent anymore, it "moved" downwards, enabling us to solve for it as we would do in any linear equation. An example will make this clear.

We deposited 20,000 $ in a bank at an annual interest rate of 5 %. The (literally million dollar) question is: how many years will it take until this grows to a million dollars? In mathematical terms this question corresponds to:

$$20,000 \cdot 1.05^x = 1,000,000$$

with x symbolizing the number of years. The first step in solving this is to bring it to the general form of an

90

exponential equation by dividing both sides by 20,000:

$1.05^x = 50$

We then apply the natural logarithm to both sides:

$ln(1.05^x) = ln(50)$

Now look at the expression on the left side. It has the same form as the left side of the logarithmic identity with a = 1.05. So we apply the identity:

$x \cdot ln(1.05) = ln(50)$

It has now transformed into a simple linear equation that we can easily solve by dividing both sides by ln(1.05):

$x = ln(50) / ln(1.05)$

Using a calculator, for example the build-in Windows calculator, we determine the required values:

$ln(50) \approx 3.91$

$ln(1.05) \approx 0.049$

Inserting this leads to:

$x \approx 80$ *years*

So with the given principal and interest rate, we would need to wait 80 years to become millionaires. Maybe not the number of years you were hoping for, but the fact that we were able to derive a number at all is thanks to the logarithmic identity.

A new population of algae has been discovered on a lake. At the time of the observation, it covered 15 m² of the 8500 m² lake and scientists were able to determine that it grows with about 8 % per week. If no measures were taken, how many months would it take for the algae population to cover the entire lake?

Again we convert this question into an equation and solve it using the identity exactly as we did above:

$15 \cdot 1.08^x = 8500$

Divide by 15 to get to the general form:

$1.08^x = 567$

Apply ln and the identity:

$\ln(1.08^x) = \ln(567)$

$x \cdot ln(1.08) = ln(567)$

Solve the linear equation for x:

$x = ln(567) \, / \, ln(1.08)$

$x \approx 82 \text{ weeks} \approx 21 \text{ months}$

As you can see, the process of solving these kinds of equations is always the same. There's only one way and it's the route via the logarithmic identity. So keep this in mind, it will enable you to solve a lot of very interesting problems.

- **Living in Harmony:**

We will start this section by looking at the harmonic series. Its name comes from an application in acoustics regarding the overtones of musical instruments. The harmonic series H(n) is defined as the sum of all reciprocals of natural numbers up to a certain number n. In mathematical terms:

$$H(n) = 1 + 1/2 + 1/3 + ... + 1/n$$

For example:

$$H(4) = 1 + 1/2 + 1/3 + 1/4 \approx 2.08$$

Simple as that. At first this sum seems like a rather artificial construct, but it does appear in a surprising amount of real-world applications. So determining this sum can be very helpful. And it was quite easy for H(4), but imagine having to evaluate H(100) or H(1000). For the latter you would need to sum 1000 numbers and then do it all over again to double-check. Not very practical.

Luckily, there's a neat approximation formula for just this sum. The higher the number n, the better the estimate will be. It is mathematically proven that as n grows to infinity, the approximation formula converges to the true value. Here it is:

$$H(n) \approx \ln(n) + 0.58$$

The value 0.58 comes from rounding off the Euler-Mascheroni constant, which should be where the 0.58 is now. But since we just want to approximate, there's no need to be overly precise. For our purposes the rounded off value will do just fine.

Imagine you are collecting stickers and the full set of the stickers consists of N different pieces. How many stickers will you most likely need to buy to complete the set? We will denote the required number of purchases by P. The solution is:

$P = N \cdot (1 + 1/2 + 1/3 + ... + 1/N)$

For a large sticker set, evaluating this expression would be rather annoying and time-consuming. Time to apply the approximation:

$P \approx N \cdot (ln(N) + 0.58)$

This looks much better. For a set of N = 50 stickers, this is how many stickers you need to buy to complete it:

$P \approx 50 \cdot (ln(50) + 0.58) \approx 225$

It's interesting to note that 50 of the 225 purchases will be for acquiring the last sticker and 25 for the one before that.

Another very cool application of the harmonic series can be found on the plus.maths.org. Imagine we start recording the daily amount of rainfall. How often can we expect weather records to be broken?

Obviously the first day will be a weather record. On the second day there's a fifty-fifty chance that there will be a new record. The expected number of weather records up to this point is:

$r(2) = 1 + 1/2$

On the third day there's a 1 in 3 chance that we will see a new weather record, leading to this expected number of records:

r(3) = 1 + 1/2 + 1/3

The pattern is now obvious. After n days of continuous weather recording, this is how many record days we can expect to see:

r(n) = 1 + 1/2 + 1/3 + ... + 1/n ≈ ln(n) + 0.58

Ten years correspond to about 3650 days. In this time the weather record will most likely be broken 9 times. In one hundred years, or 36500 days, we should see 11 record days. Note the painfully slow growth. We increased the time span by a factor of ten, yet the number of record days only grew by two.

As we saw in the last example, the harmonic series grows impossibly slowly. Here are some values to convince you:

H(1000) ≈ 7.5

H(2000) ≈ 8.2

H(3000) ≈ 8.6

H(4000) ≈ 8.9

Even though we add another thousand terms at each step, the harmonic series hardly increases in value. Even worse: the growth slows down. Where will this end? Will we reach a limiting value at some point? Or will it just grow to infinity at a terribly slow pace? Mathematically it can be proven that there's no bound. It will just keep on growing and growing.

- **Geometric Series:**

This one's a real beauty and very useful on top of that. We noted that the harmonic series featured in the last section does not converge, meaning that it does not grow to a limiting value as we include more and more terms. It simply keeps growing to infinity, which seems logical since we add an infinite number of terms. However, even with an infinite number of terms a sum can approach a limit. This is the case in the infinite geometric series.

Let's look at one example before stating the formula. Suppose we want to compute this sum:

$1 + (0.8) + (0.8)^2 + (0.8)^3 + ...$

The sum of the ...

... first 10 terms is: 4.4631

... first 20 terms is: 4.9423

... first 30 terms is: 4.9938

... first 40 terms is: 4.9993

It seems that rather than just growing and growing, the sum approaches the limiting value five. With the formula for the infinite geometric series, we can prove that. We are given a certain number x that is between zero and one. To compute the corresponding geometric sum we can use this formula:

$$1 + x + x^2 + x^3 + ... = 1 / (1 - x)$$

In the case of $x = 0.8$ we get:

$1 + (0.8) + (0.8)^2 + (0.8)^3 + ... = 1 / 0.2 = 5$

As expected. Now it might seem somewhat useless to you to have a formula for such sums. Are there actually applications for those? Plenty. As in the case of the harmonic series, the geometric series pops up surprisingly often when solving physics or math problems. It is one of these formulas most physicists and mathematicians know by heart because they need to use it over and over again.

Let's turn to some examples.

We let a ball drop from 1 m height. After each impact, it bounces back to 60 % of its previous height. What distance will the ball travel in total?

After the first impact it will rise to 0.6 m height, after the second impact to $0.6 \cdot 0.6 = 0.6^2$ m, after the third impact to $0.6 \cdot 0.6 \cdot 0.6 = 0.6^3$ m, and so on. The total distance traveled is thus (note the factor 2 since the ball rises to and drops from the computed height except for the initial drop):

$$d = 1 + 2 \cdot 0.6 + 2 \cdot 0.6^2 + 2 \cdot 0.6^3 + \dots$$

Let's rewrite this by factoring out $2 \cdot 0.6$:

$$d = 1 + 2 \cdot 0.6 \cdot (1 + 0.6 + 0.6^2 + \dots)$$

Clearly, the expression in the bracket is a geometric series with $x = 0.6$. Thanks to the formula we can compute it:

$$1 + 0.6 + 0.6^2 + \dots = 1 / 0.4 = 2.5$$

Thus the total distance traveled is:

$$d = 1 + 2 \cdot 0.6 \cdot 2.5 = 4 \ m$$

A patient with an infection is advised to take a 50 mg antibiotics tablet every day. After one day, only 15 % of the amount taken in by a tablet will remain in the body. What amount of antibiotics will be in the patient's body in the long run?

On the second day of the treatment, the amount A of antibiotics in the body will 50 mg from today's tablet and 0.15 · 50 mg from yesterday's tablet:

$A = 50 + 0.15 \cdot 50$

On the third day, we will again have 50 mg from today's tablet, 0.15 · 50 mg from yesterday's and 0.15 · 0.15 · 50 mg from the tablet taken on the first day:

$A = 50 + 0.15 \cdot 50 + 0.15^2 \cdot 50$

Continuing this train of thought, we can conclude that in the long run the amount of antibiotics in the body will be:

$A = 50 + 0.15 \cdot 50 + 0.15^2 \cdot 50 + 0.15^3 \cdot 50 + ...$

$= 50 \cdot (1 + 0.15 + 0.15^2 + 0.15^3 + ...)$

The sum in the bracket is an infinite geometric series with x = 0.15 and we compute its value from the formula:

$1 + 0.15 + 0.15^2 + 0.15^3 + ... = 1 / 0.85 \approx 1.18$

All that's left is inserting this for the bracket:

$A \approx 50 \cdot 1.18 \approx 60$

So in the long run this treatment will lead to 60 mg of antibiotics being present in the body, 50 mg from today's tablet and 10 mg rest from all the previous tablets.

Is 0.999... equal to 1? One could argue over this for hours and hours. But instead of that, we'll just calculate it. Note that we can rewrite 0.999... as such:

$0.999... = 0.9 + 0.09 + 0.009 + 0.0009 + ...$

$= 9/10 + 9/100 + 9/1000 + 9/10000 + ...$

$= 9/10 \cdot (1 + 1/10 + 1/100 + 1/1000 + ...)$

$= 9/10 \cdot (1 + 1/10 + (1/10)^2 + (1/10)^3 + ...)$

As you can see, the expression in the brackets is an infinite geometric series with x = 1/10. Let's focus on this sum:

$1 + 1/10 + (1/10)^2 + (1/10)^3 + ...$

$= 1 / (1 - 1/10) = 1 / (9/10) = 10/9$

Inserting this for the bracket we get:

$0.999... = 9/10 \cdot 10/9 = 1$

The proof is completed and 0.999... is indeed and undeniably equal to 1. Pure mathematics can be quite interesting.

I hope these examples were helpful in understanding how the geometric series arises and how we can quickly compute it.

So don't underestimate the usefulness of this formula and the geometric series. It pops up in places where you would expect it the least. As a final treat, here's an image with a simple yet brilliant proof of the formula. Enjoy!

$$s = 1 + x + x^2 + \ldots$$

$$x \cdot s = x + x^2 + x^3 + \ldots$$

$$s - x \cdot s = 1 + x + x^2 + \ldots - x - x^2 - x^3 - \ldots = 1$$

$$s \cdot (1 - x) = 1$$

$$s = 1 / (1 - x)$$

- **Poisson Distribution:**

This is an excerpt from "Statistical Snacks" by Metin Bektas.

The Poisson distribution is a discrete probability distribution, similar to the binomial distribution. One big difference though is that instead of having probabilities as inputs, we rather look at the average rate of a certain event occurring. For example, instead of being given the chance of a goal occurring during a game, we are given the average number of goals per game and go from that. This actually makes things a lot easier.

Assume we know from looking at a certain soccer team's history that it produces goals with a mean rate of 2.4 goals per game. Now we want to know how likely it is that during a particular game it will not shoot any goal. Using the Poisson distribution we can answer this question (and many more questions of this kind) straightforward:

p(no goal) = 9 %

Here's the general formula to solve such problems. We are given an average rate λ at which an event is occurring over a certain time span (goals per game, accidents per year, mails per day). If the occurrence of the event is random and independent of any previous occurrences, we can use this formula to calculate the chance that it will occur k times during said time span:

p(k occurrences) = $e^{-\lambda} \cdot \lambda^k / k!$

You are probably wondering about the exclamation mark. What does it mean to have a number followed by an exclamation mark? We call k! a factorial and read "k factorial". Whenever we see this, we just multiply all numbers down to one. For example: $3! = 3 \cdot 2 \cdot 1 = 6$ or $5! = 5 \cdot 4 \cdot 3 \cdot 2 \cdot 1 = 120$. So nothing to worry about. Of course for $0!$ this doesn't work, it is defined as $0! = 1$. Keep that in mind.

Going back to the introductory example, we wanted to know how likely it is for $k = 0$ goals to occur during a game when the average rate is $\lambda = 2.4$ goals per game:

$p(no\ goal) = e^{-2.4} \cdot 2.4^0 / 0! = 0.09 = 9\ \%$

Statistics show that in the US state of New York there are on average five tornadoes per year. How likely is it that during one year only two tornadoes will form? What's the probability of more than five tornadoes occurring?

Let's turn to the first question. All we need as inputs for the Poisson distribution is the average rate, in this case $\lambda = 5$, and the number of occurrences, in this case $k = 2$. Plugging that into the formula gives us:

$p(2\ tornadoes) = e^{-5} \cdot 5^2 / 2! = 8.4\ \%$

So the chance of only two tornadoes forming over a year is about 1 in 12. This was the simpler of the two questions. What about the chance of having more than five tornadoes? Since the Poisson distribution is infinite, we shouldn't try to do this sum:

$p(6\ tornadoes) + p(7\ tornadoes) + p(8\ tornadoes) + ...$

A better approach is to compute how likely it is to have five or less tornadoes. One minus whatever we get there is the probability of having more than five tornadoes. Let's calculate the odds of five or less tornadoes occurring by simply adding the chances for no tornado, for one tornado, and so on up to five:

$p(no\ tornado) = e^{-5} \cdot 5^0 / 0! = 0.007$

$p(1\ tornado) = e^{-5} \cdot 5^1 / 1! = 0.034$

Continuing this path until we get to five and summing all the terms results in:

$p(5\ or\ less\ tornadoes) = 0.616$

Since the probability for five or less tornadoes and the probability for more than five tornadoes must add up to one, we can quickly get the desired result:

$p(more\ than\ 5\ tornadoes) = 0.384 = 38.4\ \%$

You can easily find online calculators that do all the computing for you. I recommend the "Stat Trek Poisson Distribution Calculator", which is easy to use and also displays cumulative probabilities. This can be very helpful when answering questions featuring the phrases "at least" or "more than".

Part III: Economics

- **Inflation:**

This is an excerpt from "Business Math Basics - Practical and Simple" by Metin Bektas.

There's no denying it: things get more expensive. This happens in all economies and almost every year. At moderate rates, this increase in price level is not alarming. The picture below shows the inflation rates for the US from 1991 to 2012. Only in 2009, shortly after the financial crisis, did prices actually fall.

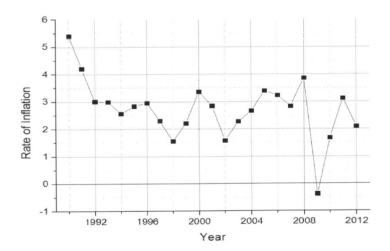

What reasons are there for inflation to occur? One way of answering this question is to take the monetarist approach and focus on the so called Equation of Exchange. It will help us to easily identify the culprit.

Let's look at the quantities necessary to understand this equation step by step and using an example. One quantity is the money supply M. It's simply the total amount of money present in the economy. For introductory purposes, I'll set this value to M = 100 billion $.

Also important is the velocity of money V. It tells us, how often each dollar (bill) is used over the course of a year. This quantity depends on the saving habits of the people in the economy. If they are keen on saving, the bills will only pass through a few hands each year, thus V is small. On the other hand, if people love to spend the money they have, any bill will see a lot of different owners, so V is large. For the introductory example, we'll set V = 5.

Note that the product of these two quantities is the total spending in the economy. If there are M = 100 billion $ in the economy and each dollar is spend V = 5 times per year, the total annual spending must be M · V = 500 billion $. This conclusion is vital for understanding the Equation of Exchange.

There are two more quantities we need to look at, one of which is the price level P. It tells us the average price of a good in the economy. If there's inflation, this is the quantity that will increase. Let's assume that in our fictitious economy the average price of a good is P = 25 $.

Last but not least, there's the number of transactions T, which is just the total number of goods sold over the entire year. We'll fix this to T = 200 billion for now and make another very important conclusion.

The product of these last two quantities is the total sales revenue in the economy. If the average price of a good is P = 25 $ and there are T = 200 billion goods sold in a year, the total sales revenue must be P · T = 500 billion $. It is no accident that the total sales revenue equals the total spending. Rather, this equality is the (reasonable) foundation of the Equation of Exchange. For the total spending to equal the total sales revenue, this equation must hold true:

$$M \cdot V = P \cdot T$$

which is just the Equation of Exchange. Now think about what will happen if we increase the money supply M in the economy, for example by printing money or government spending. We'll assume that the spending habits of the people remain unchanged (constant V). Since we increased the left side of the equation, the total spending, the right side of the equation, the total sales revenue, must increase as well.

One way this can happen is via an increase in price level P (inflation). Indeed empirical evidence shows that in the US every increase in money supply was followed by a rise in inflation later on.

Luckily there's another quantity on the right side which can absorb some of the growth in money supply. A rise in the number of transactions T (increased economic activity) following the "money shower" will dampen the resulting inflationary drive. On the other hand, a combination of more money and less economic activity can lead to a dangerous, Weimar-style hyperinflation.

At some point in your life, you probably thought to yourself: If governments can print money, why the hell don't they just make everyone a millionaire? The answer to this question is now obvious: The Equation of Exchange, that's why. If the government just started printing money like crazy, the rise in price level would just eat the newly found wealth up. Each dollar bill would gain three zeros, but you couldn't buy more with it than before.

Of course there can be much more trivial causes for inflation than a growing money supply. Prices are determined by an equilibrium of supply and demand. If demand drops, retailers have to lower their prices to sell off their stocks. Similarly, if demand suddenly increases, the retailer will be able to set higher prices, resulting in inflation. This happens for example when a new technology comes along that quickly rises in popularity. Appropriately, this kind of price level growth is called a demand-pull inflation.

- **Doubling Time / Half Life:**

Often times we deal with quantities that grow exponentially. This means that each year (month, week, ...) it changes by a fixed percentage. A typical example is compound interest. If you put 10,000 $ in a bank at an interest rate of 5 %, you will have the following amount of money in your bank account after t years:

$M = 10,000 \cdot 1.05^t$

Another example is radioactive decay. If you have 100 gram of a radioactive material that decays with 2 % per year, this is the mass that is left after t years:

$m = 100 \cdot 0.98^t$

This shows that you can easily set up an equation for future values of the quantity in this form:

$F = I \cdot (1+p)^t$

with F being the future value after t years, I being the initial value and p being the percentage change expressed in decimal numbers. In case of growth p is positive, in case of decline negative. You should keep this approach in mind, it often comes in handy.

One characteristic property of exponential growth or decline is that the time it takes for the quantity to double or halve is a constant. So if it doubles in ten years, it will double again in another ten years, double yet again during the next ten years, and so on. This doubling time (or half life in the case of decline) can be easily computed from this great formula:

$T = \ln(2) / \ln(1+p)$

Note that the doubling time does not depend in any way on the initial value. Only the percentage change counts. As for units, the computed doubling time will be in the unit of time that the percentage change is expressed in. For example, if a quantity grows 4 % per month and we input this into the formula, the resulting doubling time will be in months. Also remember to always input the percentage as a decimal number.

A typical value for the annual inflation rate in industrialized countries is about p = 3 % = 0.03. If this remained constant, how long would it take for prices to double? We can answer this very quickly and easily:

T = ln(2) / ln(1.03) ≈ 23 years

which is about one generation. At the end of World War I the inflation rate in the US rose to about p = 20 % = 0.2. What is the corresponding doubling time?

T = ln(2) / ln(1.2) ≈ 4 years

As of 2012 the world population is at about seven billion people and grows with 1.1 % per year. According to the approach from the introduction of this section, after another t years we can expect P people to live on earth:

P = 7 · 1.011ᵗ

How long does it take mankind to double its numbers if the trend continues at this rate? Let's apply the great formula to find out:

$T = ln(2) / ln(1.011) \approx 63$ *years*

So in 2075 there would be 14 billion people on earth. However, the annual growth rate has been declining since the sixties and is expected to do so in the future as well. In 1963 the annual growth rate peaked at 2.2 %, which implies a doubling time of:

$T = ln(2) / ln(1.022) \approx 32$ *years*

Luckily for all those alive and yet to be born, the explosive growth is flattening out as we speak. So in the long run the growth seems to be logistic rather than exponential.

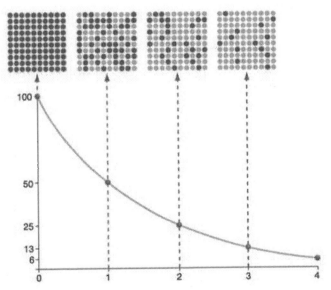

The radioactive material Polonium-210 decays at a rate of about 3.5 % per week. What is the half life of this material?

T = ln(2) / ln(1.035) ≈ 20 weeks

So if you initially had a 160 gram sample of Polonium-210, you'd be left with ...

... 80 grams after 20 weeks

... 40 grams after 40 weeks

... 20 grams after 60 weeks

... 10 grams after 80 weeks

and so on in that fashion. Where does all this mass goes? It is given off partly as a stream of alpha particles and electrons and partly as radiation, all which can be dangerous to people when exposed to this radioactive decay.

So the doubling time or half life is indeed a very useful concept that is easy to calculate on top of that. It works whenever we are faced with exponential growth.

- **Optimal Price:**

Choosing a price for a product is never easy. You can go for low prices and thus a high sales rate, but still end up with little revenue because your margin on each sale was minimal. On the other hand, you can choose high prices to maximize your margin, but again you could end up with almost no revenue because people are not willing to buy at this price and simply turn to your competition. What to do?

As so often, the in between provides the best option. Assuming that the sales rate x declines linearly with the price p, there's an optimal price p(opt) that will result in the highest possible revenue. The great news: there's is a simple formula to compute the optimal price. The bad news: you need at least two data points to use it and it'll still be only an approximation.

But let's take it step by step. You've been selling your product for some time now at price p. During this time, the sales rate was more or less constant at x. Since you feel that things could be better, you raise or lower the price to a new value p' and observe how the market reacts. The sales rate changes to x'. With these two data points, it is possible to compute the optimal price.

Before stating the formula it is helpful to define two additional quantities that we can easily derive from the two data points. The first quantity is the percentage change in price from p to p':

$\sigma(\text{price}) = (p' - p) / p$

The second quantity is (unsurprisingly) the percentage change in sales rate from x to x':

$\sigma(sales) = (x' - x) / x$

Note that the signs of these two quantities usually differ. If we increase the price, positive $\sigma(price)$, the sales rate usually falls, negative $\sigma(sales)$. Make sure to set the signs correctly, otherwise the formula will produce a false optimal price. That said, here's the formula:

$p(opt) = 0.5 \cdot p \cdot (1 - \sigma(price) / \sigma(sales))$

We can also calculate the maximum possible sales revenue r(max) using this formula:

$r(max) = - p(opt)^2 \cdot (x' - x) / (p' - p)$

Let's apply the formulas.

A company that manufactures and sells external hard drives determined that at p = 60 $ it sells x = 1350 hard drives per month and at p' = 80 $ it sells x' = 1050. Estimate the optimal price and the maximum possible sales revenue.

Let's first compute the percentage changes:

$\sigma(price) = (80 - 60) / 60 = 0.33 = 33 \%$

$\sigma(sales) = (1050 - 1350) / 1350 = -0.22 = -22 \%$

So the company increases prices by 33 % and as a reaction to that, the sales rate dropped by 22 %. We can now use the formula for the optimal price:

$p(opt) = 0.5 \cdot 60 \$ \cdot (1 - 0.33 / (-0.22))$

$p(opt) = 75 \$$

So, assuming the relationship between price and sales rate to be linear, the optimal price for the product is at 75 $. At this price the company will make this maximum possible revenue:

r(max) = -75² · (1050 - 1350) / (80 - 60)

r(max) ≈ 84,400 $ per month

This of course also means that at the optimum price, the company will sell r(max) / p(opt) ≈ 1130 hard drives per month. In the image below you can see theoretical variation of sales rate and revenue with price for this product.

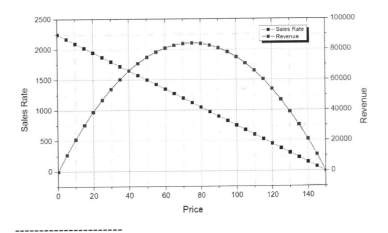

Note that the ratio σ(price) / σ(sales) determines how the optimal price p(opt) relates to the initially chosen price p. If the percentage change in prices is greater then the following percentage change in sales rate, as it was in the example, then the optimal price is greater than the initial price. If both percentage changes turn out to be equal, the optimal price coincides with the initial price.

- **Annuity:**

When you borrow a large amount of money from the bank, for example to buy an expensive car or a house, you will usually pay it back in monthly installments. These include the interest that is to be paid on the credit. This section focuses on the formula that allows you to compute the monthly rates.

What inputs do we need? Obviously we will need the principal P (in $ or any other currency), that is, the amount of money that we borrowed, and the interest rate i (expressed in decimals). Additionally, we need to know the total duration of the loan t (in years). Given these, we can calculate the annuity A, which is the annual installment.

$$A = P \cdot i \cdot (1 + i)^t / ((1 + i)^t - 1)$$

We just divide the annuity by twelve to get the monthly installment. Note that the actual value can be a few percentages higher or lower, depending on the specific fees and conditions.

Granted, the formula does not look very appealing. But don't be intimidated by it. As long as you input the right values and calculate carefully, nothing will go wrong. It goes without saying that the formula is of great importance. It is one of the most often used formulas in banking and will become relevant to almost all of us at some point in our lives.

We want to get a P = 200,000 $ loan from the bank and pay it back over the next t = 20 years. The bank agrees to loan *us this sum at an interest rate of i = 4 % = 0.04. What will the monthly installments be? To find that out, we simply plug in all we know into the annuity formula:*

A = 200,000 · 0.04 · 1.04^{20} / (1.04^{20} -1)

≈ 17,529 / 1.19 ≈ 14,730 $

So the monthly installments will be 14,730 $ / 12 ≈ 1230 $.

The computed monthly rate turns out to be too high for us. We would like to reduce the installments by increasing the duration of the loan to t = 30 years. The bank agrees. How does this affect the monthly rate?

A = 200,000 · 0.04 · 1.04^{30} / (1.04^{30} -1)

≈ 25,947 / 2.24 ≈ 11,580 $

which translates into a monthly installment of 965 $. So the ten additional years of responsibility reduced the rate by about 20 %. Is that worth it? It's your call.

The annuity formula has a very useful inversion. Sometimes we already know what monthly rate (and thus annuity A) we would like to or are able to pay. Given the principal P and interest rate i, we can then compute the duration t of the loan. To do that, we first calculate this quantity:

x = A/S - i

and insert the result into this equation:

$$t = \ln(1 + i / x) / \ln(1 + i)$$

Let's go back to the P = 200,000 $ loan at i = 4 % interest rate. We would like to have a monthly rate of 1100 $. What is the corresponding duration of the loan? Note that this monthly rate corresponds to an annuity of 1100 $ · 12 = 13,200 $.

First we compute the mysterious x:

x = 13,200 / 200,000 - 0.04 ≈ 0.026

Now we can apply the formula:

t = ln(1 + 0.04 / 0.026) / ln(1.04) ≈ 24 years

Let's make this proposal and hope that the bank is cooperative. Knowing the formulas certainly didn't hurt. It gets you on a level playing field with your bank.

The annuity formula or its inversion is certainly no smooth sailing and don't even bother to memorize it. But its difficulty is overrated. In the end, it's just another formula to plug values into. You don't need to be a trained economist to do that.

- **Queues:**

Nobody likes waiting in line. Still we are forced to do just that almost every day: at the bank, at the doctor's office, at the fast food restaurant, at the gas station, ... In this section we will take a closer look at waiting in single line, multiple channel systems, meaning that in such cases there's one line for waiting customers and one or more servers (see image).

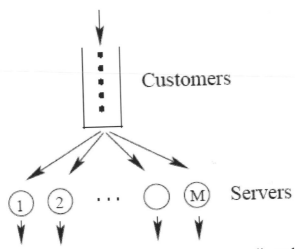

The formulas for this case are very complicated and extremely useful. We need three quantities as inputs: the arrival rate λ (in customers per unit time), the service rate μ (also in customers per unit time) and the number of servers M. Let's focus on the case of having only one server first, M = 1. The average waiting time per customer T (in the given unit time) will be:

$$T = \lambda / (\mu \cdot (\mu - \lambda))$$

With this computed, we can also easily state the average number of customers N in the queue:

$$N = \lambda \cdot T$$

Let's do an example for these relatively simple formulas before moving on to the really brutal stuff.

At the doctor's office the patients arrive at a rate of $\lambda = 5$ patients per hour. The doctor can serve $\mu = 6$ patients per hour. What will be the average waiting time for a patient? How many people will be in the waiting room on average?

$T = 5 / (6 \cdot 1) = 0.83$ hours $= 50$ minutes

$N = 5 \cdot 0.83 \approx 4$ people in waiting room

Now let's turn to the case of more than one server. Before we can compute the waiting time, we will need to evaluate the probability p of having no customers in the system:

$$1/p = \text{sum from n=0 to n=M-1 over } ((\lambda/\mu)^n / n!)$$

$$+ (\lambda/\mu)^M \cdot M \cdot \mu / (M \cdot \mu - \lambda) \cdot 1 / M!$$

I told you it's gonna be brutal! If you're wondering about the exclamation mark, take a look at the section "Poisson Distribution". You'll find an explanation there.

For the special case of having $M = 2$ servers, the overly complicated formula reduces to this more pleasing one:

$$1/p = 1 + \lambda/\mu + (\lambda/\mu)^2 \cdot 2 \cdot \mu / (2 \cdot \mu - \lambda) \cdot 0.5$$

Once we've calculated p, we can use this formula to derive the waiting time per customer T.

$$T = p \cdot \mu \cdot (\lambda/\mu)^M / ((M-1)! \cdot (M \cdot \mu - \lambda)^2)$$

Again we can simplify this for M = 2 servers:

$$T = p \cdot \mu \cdot (\lambda/\mu)^2 / (2 \cdot \mu - \lambda)^2$$

Luckily, the handy formula for computing the average number of customers N in the queue remains: $N = \lambda \cdot T$. That being said, let's turn an example.

We will stick said doctor's office with an arrival rate $\lambda = 5$ and service rate $\mu = 6$ patients per hour. The doctor rightly feels that 50 minutes waiting time is too much and invites a colleague to join him. So now there are M = 2 servers. How does this impact the waiting time and number of patients in the waiting room?

First we need the probability of having no patient in the system (no patient in the waiting room or being served). We can use the simplified formula:

$$1/p = 1 + 5/6 + (5/6)^2 \cdot 2 \cdot 6 / (2 \cdot 6 - 5) \cdot 0.5$$

$$1/p \approx 1 + 0.83 + 0.60 = 2.43$$

$$p = 0.41$$

Now we can compute the average waiting time T and the average number of people in the waiting room:

$$T = 0.41 \cdot 6 \cdot (5/6)^2 / (2 \cdot 6 - 5)^2$$

T ≈ 0.03 hours ≈ 2 minutes

N = 5 · 0.03 ≈ 0 people in waiting room

So the additional server made a huge difference, reducing the waiting time from 50 minutes to a mere 2 minutes and effectively emptying the waiting room.

Note that all the formulas only work when $M \cdot \mu$ is greater than λ and the customers are served by the FIFO (First In, First Out) principle. Also it was assumed that no customer leaves the queue before being served. If this occurs regularly, then the average queue size will be shorter than the computed value.

- **Risky Games:**

Whenever we do business, there's a chance of success and a chance of failure. How it will turn out depends on many things: our skills, our business partners, the market situation, political decisions, and so on. A simple formula from statistics can help us deal with risk by allowing us to compute the expected value of the gains or losses.

It is all based on the concept of probability distributions, so we need to take a look at those first. A (discrete) probability distribution lists all possible outcomes along with their probability. For example, imagine we are offered a game of dice. We put in a wager of 5 $. If we roll a six, we get the wager back plus 20 $, if we don't, we lose the wager. Since we roll a six with the probability 1/6, here's the probability table for this game:

- -5 $ with the probability 5/6

- 20 $ with the probability 1/6

Note that in such distributions the probabilities must always add up to one. So is it worth playing this game? How much is our expected gain or loss per round? This can be answered using the following great formula. Assume we are given a probability distribution with the numerical outcomes $n(1)$, $n(2)$, $n(3)$, ... and their respective probabilities $p(1)$, $p(2)$, $p(3)$, ... The expected outcome per round is:

$$e = n(1) \cdot p(1) + n(2) \cdot p(2) + n(2) \cdot p(2) + \ldots$$

Thus, we simply multiply all the numerical outcomes with their respective probabilities and do the sum. Let's compute the expected value for our game of dice:

$e = -5\ \$ \cdot 5/6 + 20\ \$ \cdot 1/6 \approx -0.83\ \$$ per round

So this game is not at all favorable to us, in the long run we can only lose. If we play 100 rounds, we can be expected to lose 83 $. It's better to turn down this offer and wait for a better one to come along.

By the way, with the concept of the expected value it is very simple to define what a fair game is. If e = 0, the game is fair. In the above situation a fair pay-out in case of rolling a six would have been 25 $. With this value we get:

$e = -5\ \$ \cdot 5/6 + 25\ \$ \cdot 1/6 = 0\ \$$ per round

Here neither the player nor the casino is favored. Let's turn to some examples now.

A start-up business wants to borrow 100,000 $ from a bank at an interest rate of 6 %. The probability of default is estimated to be 8 %. Should the bank agree and go ahead with the credit? To answer that, let's take a look at the probability distribution. In the case of the loan being paid back, the bank gains $0.06 \cdot 100,000\ \$ = 6000\ \$$. However, if the start-up business defaults, the bank will have a loss of 100,000 $.

- *-100,000 $ with the probability 0.08*

- *6000 $ with the probability 0.92*

Let's look at the expected value, that is, the expected gain or loss per credit of this type.

$e = -100,000\ \$ \cdot 0.08 + 6000\ \$ \cdot 0.92$

$\approx -2480\ \$\ per\ credit$

So for the bank this set up is not favorable, the interest rate is too low to make up for the high chance of default. Here you can see why coupling interest rates to risk makes sense.

Again the start-up business wants to borrow 100,000 $ from a bank and the probability of default is 8 %. How should the bank set the interest rate i in order to have an expected gain of 1000 $ per credit of this type?

Again, let's look at the probability distribution. The bank gains $i \cdot 100,000$ $ if the loan is paid back and loses 100,000 $ if the start-up business defaults.

- -100,000 $ with the probability 0.08

- $i \cdot 100,000$ $ with the probability 0.92

Since we want to have an expected value of 1000 $, we can use the great formula to set up and solve this equation:

$1000 = -100,000 \cdot 0.08 + i \cdot 100,000 \cdot 0.92$

$1000 = -8000 + i \cdot 92,000$

$9000 = i \cdot 92,000$

$i \approx 0.1 = 10\ \%$

It should be noted that the expected value is a number that is approached in the long run. In the dice game from the introductory example you could indeed make a profit despite its negative expected value. If you played only two rounds and you won both, you would have gained 40 $.

The expected value tells you what balance is most likely to occur after a large number of rounds and this very efficiently. It is mathematically proven (Law of Large Numbers) that as the number of rounds grows, the actual balance will converge to the computed value, no exception.

The image below shows the difference between the actual profit per round for a dice game resulting from a randomized computer simulation and the theoretical profit per round as computed from the formula. The convergence is clearly visible. The more rounds we play, the smaller this difference becomes. It's the Law of large numbers in action.

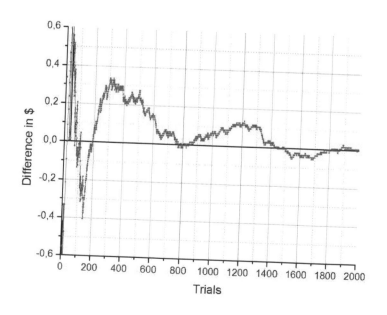

Part IV: Appendix

- **Unit Conversions**

Since we often need to convert units from the United States customary system (USCS) to the metric (SI) system and vice versa, here's a list of commonly needed conversion factors.

Lengths, SI to USCS:

Multiply meters with 3.28 to get to feet

- Multiply meters with 1.09 to get to yards

- Multiply meters with 0.00062 to get to miles

- Multiply kilometers with 3281 to get to feet

- Multiply kilometers with 1094 to get to yards

- Multiply kilometers with 0.62 to get to miles

Lengths, USCS to SI:

- Multiply feet with 0.30 to get to meters

- Multiply feet with 0.00030 to get to kilometers

- Multiply yards with 0.91 to get to meters

- Multiply yards with 0.00091 to get to kilometers

- Multiply miles with 1609 to get to meters

- Multiply miles with 1.61 to get to kilometers

To convert a squared to a squared unit, use the square of the conversion factor. For example you multiply m^2 by $3.3^2 \approx 10.9$ to get to ft^2. In a similar fashion, using the cube of the conversion factor, you can convert cubed units.

Speeds:

- Multiply m/s with 3.6 to get to km/h

- Multiply m/s with 2.23 to get to mph

- Multiply km/h with 0.28 to get to m/s

- Multiply km/h with 0.62 to get to mph

- Multiply mph with 0.45 to get to m/s

- Multiply mph with 1.61 to get to km/h

Other commonly used units:

- Multiply pounds with 0.45 to get to kilograms

- Multiply kilograms with 2.22 to get to pounds

- 1 liter = 0.001 m^3

- Multiply liters with 0.62 to get to gallons

- Multiply gallons with 3.79 to get to liters

- Celsius to Fahrenheit: $°F = 1.8 \cdot °C + 32$

- Fahrenheit to Celsius: $°C = 5/9 \cdot (°F - 32)$

- Kelvin to Celsius: $°C = K - 273.15$

- Celsius to Kelvin: $K = °C + 273.15$

- **Unit Prefixes**

peta (P) = 1,000,000,000,000,000 = 10^{15}

tera (T) = 1,000,000,000,000 = 10^{12}

giga (G) = 1,000,000,000 = 10^{9}

mega (M) = 1,000,000 = 10^{6}

kilo (k) = 1,000 = 10^{3}

deci (d) = 0.1 = 10^{-1}

centi (c) = 0.01 = 10^{-2}

milli (m) = 0.001 = 10^{-3}

micro (μ) = 0.000,001 = 10^{-6}

nano (n) = 0.000,000,001 = 10^{-9}

pico (p) = 0.000,000,000,001 = 10^{-12}

femto (f) = 0.000,000,000,000,001 = 10^{-15}

64286991R00086

Made in the USA
Columbia, SC
12 July 2019